John Mulcahy

Principles of Modern Geometry

With numerious applications to plane and spherical figures; and an appendix, containing questions for exercise. Intended chiefly for the use of junior students.

Second Edition

John Mulcahy

Principles of Modern Geometry
With numerous applications to plane and spherical figures; and an appendix, containing questions for exercise. Intended chiefly for the use of junior students. Second Edition

ISBN/EAN: 9783337312367

Printed in Europe, USA, Canada, Australia, Japan

Cover: Foto ©berggeist007 / pixelio.de

More available books at **www.hansebooks.com**

PRINCIPLES

OF

MODERN GEOMETRY,

WITH NUMEROUS APPLICATIONS TO

PLANE AND SPHERICAL FIGURES;

AND

AN APPENDIX,

CONTAINING QUESTIONS FOR EXERCISE.

INTENDED CHIEFLY FOR THE USE OF JUNIOR STUDENTS.

BY

JOHN MULCAHY, LL. D.,

LATE PROFESSOR OF MATHEMATICS, QUEEN'S COLLEGE, GALWAY.

Second Edition, Revised.

DUBLIN:
HODGES, SMITH & CO. 104, GRAFTON STREET,
BOOKSELLERS TO THE UNIVERSITY.

1862.

PREFACE.

The object of these pages is to lay down and illustrate the more elementary principles of those Geometrical Methods which, in recent times, have been so successfully employed to investigate the properties of figured space.

The importance of the principles in question seems to render it advisable that the student should enter on their application at an early period of his progress; and, in accordance with this view, examples in Plane and Spherical Geometry are here given in considerable numbers.

The scope and extent of the present work may be collected with tolerable accuracy from the Table of Contents; but it is necessary to state, for the information of the reader, the amount of Mathematical knowledge which he is supposed to possess. The preliminary Propositions required for the perusal of the first five Chapters are to be found, with few exceptions, in the first six Books of Euclid's Elements. Some occasional deductions, involving the formulæ of Plane Trigonometry, are appended to these Chapters in the form of Notes. In the sixth Chapter the fundamental notions of Algebraic Geometry are referred to. The seventh, eighth, ninth, and tenth Chapters presuppose an acquaintance with the ordinary principles of Spherical Trigonometry; and in the last two Chapters some of the properties of Curves of the second degree are assumed.

From this statement it will be gathered, that the work is in a great degree of a *supplementary* nature, and that the subjects embraced have some diversity of character. It has been attempted, however, to preserve throughout a certain unity of design, and a due connexion of the various parts.

Those who are acquainted with the writings of Poncelet and Chasles will readily appreciate the extent to which the author has borrowed from these distinguished geometers in the present publication. He is also indebted to the Additions contained in Professor Graves's Translation of Chasles's Memoirs of Cones and Spherical Conics; and on several occasions he has consulted with advantage Dr. Salmon's Treatise on Conic Sections. It is to be added, that many of the examples throughout the work, and of the questions in the Appendix, are taken from the Examination Papers of Trinity College, Dublin.

CONTENTS.

CHAPTER I.
HARMONIC PROPORTION AND HARMONIC PENCILS.

	PAGE
Harmonic Proportion	1
Examples on Harmonic Proportion	4
Harmonic Pencils	7
Transversals	8

CHAPTER II.
ANHARMONIC RATIO AND INVOLUTION.

Anharmonic Ratio	13
Examples on Anharmonic Ratio	15
Copolar Triangles	18
Anharmonic Properties of the Circle	19
Problems relating to Anharmonic Ratio	22
Hexagon inscribed in a Circle	24
Involution	26
Examples on Involution	29

CHAPTER III.
POLES AND POLARS IN RELATION TO A CIRCLE.

Poles and Polars	32
Polar Properties of Quadrilaterals	35
Method of Reciprocation	37
Problems relating to the Theory of Polars	44

CHAPTER IV.
THE RADICAL AXES AND CENTRES OF SIMILITUDE OF TWO CIRCLES

The Radical Axis	51
The Radical Centre	57
Centres of Similitude	59
Axes of Similitude	64

CHAPTER V.

ADDITIONAL EXAMPLES ON THE SUBJECTS CONTAINED IN THE FIRST FOUR CHAPTERS.

	PAGE
Additional Examples on Harmonic Proportion and Harmonic Pencils	66
Additional Examples on Transversals	69
Additional Examples on Anharmonic Ratio	75
Additional Examples on Involution	78
Additional Examples on the Theory of Polars	80
Additional Examples on Radical Axes and Centres of Similitude	90

CHAPTER VI.

THE PRINCIPLE OF CONTINUITY . 96

CHAPTER VII.

ELEMENTARY PRINCIPLES OF PROJECTION.

Method of Projection	109
Examples on Projection	113
Projection of Angles	116

CHAPTER VIII.

SPHERICAL PENCILS AND SPHERICAL INVOLUTION.

Anharmonic Properties of Four Planes	117
Anharmonic Ratio on the Sphere	119
Harmonic Properties on the Sphere	121
Anharmonic Properties of a Lesser Circle	127
Spherical Involution	129

CHAPTER IX.

POLAR PROPERTIES OF CIRCLES ON THE SPHERE.

Polar Properties of Lesser Circles	133
Method of Reciprocation on the Sphere	140
Supplementary Figures	142
Examples on Spherical Reciprocation	145

CHAPTER X.

RADICAL AXES AND CENTRES OF SIMILITUDE ON THE SPHERE.

The Radical Axis on the Sphere	152
The Radical Centre on the Sphere	153
Centres of Similitude of Two Lesser Circles	154
Axes of Similitude on the Sphere	158

CHAPTER XI.

PROPERTIES OF THE SPHERE CONSIDERED IN RELATION TO SPACE.

	PAGE
Poles and Polar Planes	159
Reciprocal Surfaces	162
Stereographic Projection	165
Inverse Surfaces	167

CHAPTER XII.

PROPERTIES OF PLANE AND SPHERICAL SECTIONS OF A CONE.

Plane Sections of a Cone	170
The Spherical Ellipse	172
The Spherical Hyperbola	181
Projective Properties of Plane and Spherical Conics	182
Orthographic Projection	197

APPENDIX.

Questions on Elementary Plane Geometry	205
Questions relating to Circles on the Sphere	213
Miscellaneous Questions on the foregoing Subjects	217

INDEX . . 225

PRINCIPLES OF MODERN GEOMETRY.

CHAPTER I.

HARMONIC PROPORTION AND HARMONIC PENCILS.

ART. 1. THREE quantities are said to be in harmonic proportion, when the first is to the third as the difference between the first and second is to the difference between the second and third. Thus, 3, 4, 6, are in harmonic proportion.

It follows from this definition, that the three quantities will still be in harmonic proportion when their order is inverted, or when they are altered in the same ratio. Thus 6, 4, 3, are in harmonic proportion, and so also are $3m$, $4m$, $6m$, m being any number.

Three quantities are said to be in arithmetic proportion, when the difference of the first and second is equal to the difference of the second and third.

Three quantities are said to be in geometric proportion, when the first is to the second as the second to the third; that is, when they are proportional in the sense of the Fifth Book of Euclid.

The relation between these three kinds of proportion may be thus exhibited :—

In arithmetic proportion the differences are equal; that is, they are as the *first* to *itself*. In geometric proportion they are as the first to the *second* (by conversion and alternation). And in harmonic proportion they are as the first to the *third*.

2. Let a right line AB, be cut internally at O, and externally at O' in the same ratio; that is, so that AO : BO : : AO' : BO'; then,

OO' is evidently an harmonic mean between AO' and BO'. Also (by alternation) $AO : AO' :: BO : BO'$; that is, AB is an harmonic mean between AO and AO'. In this case the whole line AO' is said to be cut harmonically. If the line AB be considered as given, the points O and O' may be conceived to vary simultaneously, and are said to be *harmonic conjugates*. In like manner A and B will be harmonic conjugates when OO' is considered as a given line.

The rectangle under AO and BO' is equal to the rectangle under BO and AO'; or, more concisely, $AO \cdot BO' = AO' \cdot BO$.

Conversely, when a line is divided into three parts, such that the rectangle under the whole line and the middle part equals the rectangle under the extreme parts, the line is cut harmonically.

3. *If from the same point C, and in the same direction, three portions,* CO, CB, CO', *be taken, so that* $CO \cdot CO' = CB^2$; *and if, in the opposite direction, CA be taken equal to CB, the whole line AO' will be cut harmonically.*

For, since $CO : CB :: CB : CO'$, it follows (by conversion and alternation) that $CB + CO : CB - CO :: CO' + CB : CO' - CB$; that is, $AO : BO :: AO' : BO'$.

Conversely. *If a line AO' be cut harmonically, and either of the harmonic means (see Art. 2), as AB, be bisected the three portions CO, CB, CO', measured from the point of bisection C, to the remaining points in the line, will be in geometric proportion.*

This readily follows from the last by reasoning *ex absurdo*. It may also be proved directly :—since $AO' : BO' :: AO : BO$, we have $AO' + BO' : AO' - BO' :: AO + BO : AO - BO$; that is, twice CO' : twice CB :: twice CB : twice CO, and therefore, $CO : CB :: CB : CO'$.

Hence, *when the line AB is given, the conjugate points O, O', considered as variable, move in opposite directions, and the harmonic conjugate of the middle point C is at an infinite distance.* For, $CO \cdot CO' = CB^2 = $ a constant quantity; therefore, if CO increase, CO' must diminish, and if CO decrease indefinitely, CO' must increase indefinitely.

From the last remark, again, it follows, that *when two infi-*

nite quantities differ by a finite quantity, their ratio is one of equality. For, when O' goes to infinity, AO' and BO', both infinite, have a finite difference, AB, and are in the ratio of AC : BC, that is, in a ratio of equality.

It is scarcely requisite to observe, that *when three points of section of a line cut harmonically are given, a definite pair being conjugate, the fourth is determined.*

4. *To exhibit the relation between the arithmetic, geometric, and harmonic means between two given right lines.* Let AO' and BO' be the given extremes; on AB describe a semicircle; draw the tangent O'T and TO perpendicular to the diameter. Then CO' is evidently the arithmetic mean. O'T is the geometric mean (since $AO' \cdot BO' = O'T^2$). And OO' is the harmonic mean; because, CTO' being a right-angled triangle, $CO \cdot CO' = CT^2 = CB^2$ and, therefore (Art. 3), AO' is cut harmonically. Again, since CTO' is a right-angled triangle, CO' : TO' :: TO' : OO'; that is, *the arithmetic, geometric, and harmonic means are geometrically proportional*, the arithmetic being the greatest, and the harmonic the least of the three, unless the extremes become equal, in which case the three means are also equal.

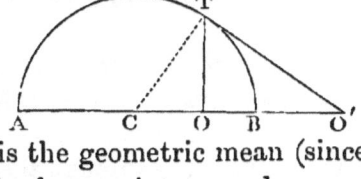

5. From the preceding construction some important properties of lines in harmonic proportion may be deduced.

1°. $CO' \cdot OO' = O'T^2 = AO' \cdot BO'$; therefore (doubling both of the equal quantities), $(AO' + BO') \cdot OO' = 2AO' \cdot BO'$; that is, *the rectangle under the harmonic mean and the sum of the extremes equals twice the rectangle under the extremes.*

2°. $2CO \cdot OO' = 2OT^2$ (since CTO' is a right-angled triangle); therefore $(AO - BO) \cdot OO' = 2OT^2 = 2AO \cdot BO$ (Lardner's Euclid, B. ii. Prop. 14); that is, *the rectangle under the harmonic mean and the difference of the differences* (see Art. 1) *equals twice the rectangle under the differences.**

---

* Considered algebraically, the two properties just given may be included in one enunciation. For, the equation in the second may be written $(OB - OA) \cdot OO' = 2OB \cdot - OA)$. Now, in applying algebra to geometry, a change of direction corresponds to a change of sign. Taking this into account, the two properties alluded to may be

3°. By the *reciprocal* $\left(\dfrac{1}{l}\right)$ of a line (l) is meant a third proportional to that line and some given line taken as a linear unit; thus, taking O'T as the unit, O'A, O'C, O'B are the reciprocals of O'B, O'O, O'A; that is, *the reciprocals of three harmonicals are in arithmetic proportion.* Also, taking OT as the unit, we have

$$\frac{1}{\mathrm{OO'}} = \frac{1}{2}\left(\frac{1}{\mathrm{OB}} - \frac{1}{\mathrm{OA}}\right).$$

EXAMPLES ON HARMONIC PROPORTION.

6. We shall now give a few examples on the preceding principles.

1°. When two circles cut each other *orthogonally* (that is, so that tangents, at a point of intersection, are at right-angles), any line AO' passing through the centre, C, of either, and meeting the other, is cut harmonically. For, in this case, it is evident that the tangent to one circle at the point 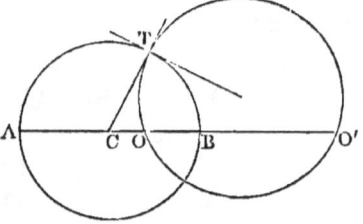 of intersection, T, passes through the centre, C, of the other; therefore, $CT^2 = CO \cdot CO'$; but $CB = CT$; therefore (Art. 3) AO' is cut harmonically.

Conversely, if a line is cut harmonically, any circle passing through one pair of conjugates is cut orthogonally by the circle whose diameter is the distance between the other pair.

2°. Suppose it were required *to draw from two given points,* A, B, *two right lines,* AT, BT, *to a certain point* T *in a given curve, so that the ratio of* AT *to* BT *should be a maximum.*

Join the given points, and let TO and TO' be the bisectors of the angle ATB and its supplement; then (Lardner's Euclid, B. vi. Prop. 3) AT : BT :: AO : BO :: AO' : BO'. Now, AO : BO is a maximum ratio when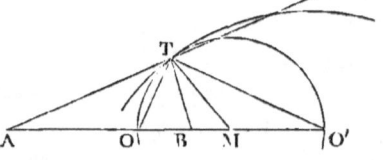

---

thus expressed: "When a line is cut harmonically, the *distance from* any of the points of section (taken as origin) *to its conjugate*, equals twice the product of the *distances* from the same point to the other pair of conjugates divided by their *sum*."

O is the nearest possible to B; that is (since the conjugate points O, O' move in opposite directions (Art. 3)) when OO' is a minimum or maximum; therefore, as OTO' is a right angle, the problem proposed would be solved if we could find, in the given line, AB, a pair of harmonic conjugates, O, O', such that a semicircle described on OO' should just touch the given curve.

To reduce the question still further, let M be the middle point of OO'; then (Art. 3) $AM \cdot BM = MO^2 = MT^2$; and therefore, a circle through A, B, T would touch MT at the point T. Hence, *this circle must be orthogonal to the given curve at the required point*, since MT is at right angles to the line touching the curve at that point.

The solution cannot, of course, be completed unless the particular nature of the given curve is taken into account.

If the question had been, to make the ratio AT : BT a minimum, we should find, in a similar way, that OO' must be a maximum or minimum. A circle on it as diameter must touch the curve, and the further reduction already made will still hold good.

We shall illustrate what has been said by taking a circle for the given curve.

Let N be the centre of the given circle, and let the required orthogonal circle cut the line AN in P. Join NT; then, since NT is a tangent, $AN \cdot NP = NT^2$, and, therefore, the point P is determined. This is the analysis of the question, and indicates the following construction. Join the centre

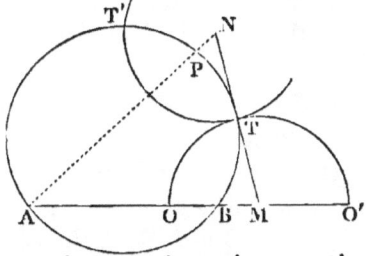

N of the given circle to one of the given points A; on the joining line take NP a third proportional to AN and the radius of the given circle; describe a circle through the points A, B, P; this circle will cut the given circle in two points, one of which (T) will give the ratio AT : BT a maximum.

The other point of intersection (T') of the two circles renders the ratio AT' : BT' a minimum.

3°. Let it now be proposed *to find a point in a given right line, the ratio of whose distances from two given points, A, B, shall be a maximum or minimum.*

At the middle point C of the line AB erect a perpendicular to meet the given line in D; with D as centre, and DB as radius, describe a circle; this circle will cut the given line in two points T, T', one of which answers to the maximum, and the other to the minimum. For, this circle 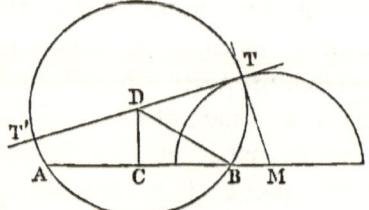 evidently cuts the given line orthogonally at T and T'; and therefore (by the reasoning in the last example) these points afford the solution of the question.

There are some particular cases of the proposition just proved which we shall leave to the consideration of the reader:—

Given in magnitude and position the base of a triangle, whose vertex lies on a given right line bisecting the base, the ratio of its sides is a maximum, when the vertical angle is right.

Given the base and one base angle (supposed acute), the ratio of the sides is a maximum, when the other base angle is right.

Given the base and area of a triangle, the ratio of the sides is a maximum, when the difference of the base angles equals a right angle.

4°. It is required to express the perpendicular on the base of a triangle, in terms of the radii of the *exscribed* circles touching the two sides. (A circle is said to be exscribed to a triangle, when it touches any of the three sides, and the productions of the other two).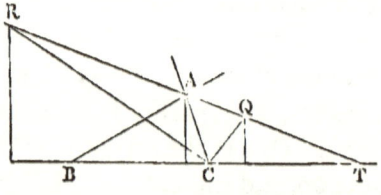

Let Q and R be the centres of the two circles, whose radii we shall express by r_2, r_3. Now, the line joining the centres evidently passes through the vertex A, and cuts the base produced in a certain point T; it is also evident that the angles ACB and ACT are bisected; therefore, (Lard. Euc. B. vi. Prop. 3) RT, AT, QT, are in harmonic proportion; and therefore so also are the three perpendiculars drawn from R, A, Q, upon the base (Art. 1). Calling the perpendicular of the triangle p, we have finally (Art. 5, 1°) $p \cdot (r_2 + r_3) = 2r_2 \cdot r_3$, and therefore $p = \dfrac{2r_2 \cdot r_3}{r_2 + r_3}$.

5°. To express the same perpendicular, in terms of the radii of the inscribed circle and the exscribed circle touching the base.

Let O and P be the centres of the two circles, whose radii we shall express by r and r_1. Now, the line joining the centres evidently passes through the vertex A, and meets the base in a certain point T; also, the line AP is cut harmonically, as the angle ACT and its supplement are bisected, and therefore (Art. 5, 2°) AT · (PT - TO)=2·PT·TO. But p, r, r_1, being perpendiculars on the base from A, O, P, are respectively proportional to AT, TO, PT; therefore $P · (r_1 - r) = 2r_1 · r$, and finally, $p = \dfrac{2r_1 · r}{r_1 - r}$.

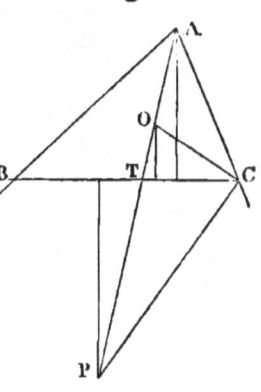

HARMONIC PENCILS.

7. "Four right lines drawn from the same point O, and cutting a right line AD harmonically in B and C (called an *harmonic pencil*), will also cut harmonically any other right line A'D' meeting them."

For, through C and C' (see fig.) draw KL and K'L' parallel to AO; then, since AD is cut harmonically, AD : DC :: AB : BC; and, therefore, by similar triangles, AO : CL :: AO : KC. Hence KC =CL, and consequent-ly, K'C'=C'L'; whence it readily follows that A'D' : C'D' :: A'B' : B'C'; that is, the line A'D' is cut harmonically.

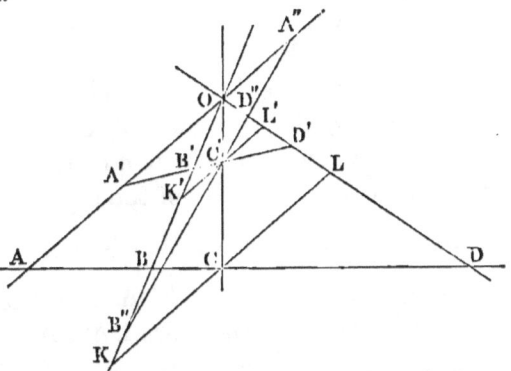

A similar proof shews that if A' be taken on the production of AO, as for instance at A", a line such as A"B" is cut harmonically.

It follows immediately from the last remark, that when all the legs of an harmonic pencil are produced through the vertex, four other harmonic pencils may be said to be formed. Strictly speaking, however (considering a right line as infinite), these with the original constitute one and the same pencil.

If B'D' coincide with K'L', the point A' is at an infinite distance, and, therefore, since K'L' is bisected, in this case also we have (Art. 3) the transversal A'D' cut harmonically.

The general property of an harmonic pencil now established was known to the ancients. (See Math. Col. of Pappus, Book vii. Prop. 145.)

The alternate legs of an harmonic pencil are said to be *conjugate*. It is evident that *when three legs of an harmonic pencil are given, a definite pair being conjugate, the fourth is determined.*

8. *If two alternate legs of an harmonic pencil be at right angles, one of them bisects the angle under the remaining pair of legs, and the other bisects the supplemental angle.*

Let the angle AOC (see last figure) be right; then, KCO is a right angle, and, as KC = CL (Art. 7), the angle BOD is bisected. Again, as AOC is a right angle, it is evident that AO bisects the supplement of BOD.

TRANSVERSALS.

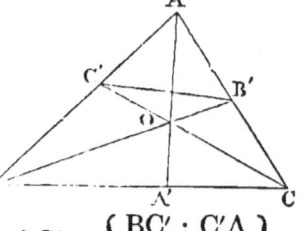

9. LEMMA 1.—If three right lines AA', BB', CC', be drawn from the angles of a triangle ABC to meet in any point O, the segments of one side of the triangle will be in a ratio compounded of the ratio of the segments of the other sides, that is, BA' : AC' :: $\left\{ \begin{array}{l} BC' : C'A \\ AB' : B'C \end{array} \right\}$.

For, BA' : A'C :: △ ABA' : △ ACA'; also, BA' : A'C :: △ BOA' : △ COA'; therefore BA' : A'C :: △ ABO : △ ACO, or :: $\left\{ \begin{array}{l} \triangle ABO : \triangle CBO \\ \triangle CBO : \triangle ACO \end{array} \right\}$, that is, :: $\left\{ \begin{array}{l} AB' : B'C \\ BC' : C'A \end{array} \right\}$. Q. E. D.

This relation may be otherwise expressed as follows:

We have (Euclid, B. vi. Prop. 23) BA' : A'C :: AB' · BC' : B'C · C'A; and, therefore, AB' · CA' · BC' = A'B · C'A · B'C, this is, *the continued products of the alternate segments of the sides are equal.*

Similar results will hold good when the point O is outside the triangle. In that case *two* of the points A', B', C' will lie on the sides produced.

Conversely, it follows, *ex absurdo*, that if the relation above-mentioned obtains, amongst the segments of the sides of a triangle made by lines drawn from the angles (under the restriction just stated, relative to a point outside), those lines will meet in one point.

LEMMA 2.—If a right line C'A' be drawn, cutting the three sides of a triangle ABC, their segments so formed will have a relation similar to that expressed in the former Lemma.

For, draw CO parallel to AB (see figs.); then $BA' : A'C :: BC' : CO ::$
$$\begin{cases} BC' : C'A \\ C'A : CO \end{cases};$$
but $C'A : CO :: AB' : B'C$; therefore $BA' : A'C ::$
$$\begin{cases} BC' : C'A \\ AB' : B'C \end{cases}.$$
And so on as before.

Conversely, if a point be taken on each side of a triangle (or the side produced), so that the segments thus formed shall satisfy the relation above referred to, those three points will be in one right line, provided that an *odd* number of the points shall lie on the productions of the sides.

The propositions above given are of great importance in the *theory of transversals*. They are particular cases of more general propositions, which shall be given in the sequel.

These Lemmas being premised, we shall now explain the harmonic properties of a triangle.

" Let three right lines AA', BB', CC', be drawn (as in Lemma 1) from the angles of a triangle, to meet in a point O, and let the lines A'B', B'C', C'A' be produced to meet the sides AB, BC, CA, respectively, in C''', A'', B''; then, 1°, all the lines on the figure so formed are cut harmonically, and, 2°, the points A", B", C", are in one right line."

1°. By the foregoing Lemmas BA' : A'C :: BA" : CA"; that is, BA" is cut harmonically. A similar proof applies to BC" and AB". Again, if AA" be joined, as BA" is cut harmonically, so also is C'A" (Art. 7); and in like manner C"A' and B"C'. For similar reasons BB' is cut harmonically, and also AA' and CC'.

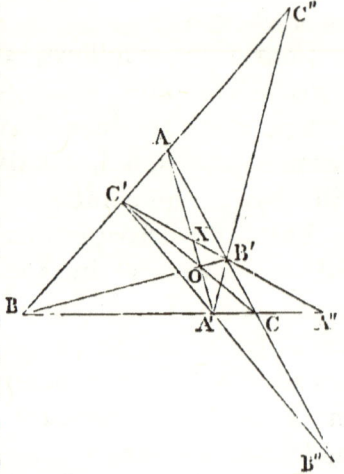

2°. Join BB"; now, if the three lines meeting at B" be taken as three legs of an harmonic pencil (B"B and B"C being supposed conjugate), the fourth is determined (Art. 7); but it must pass through A" and C", since these points are harmonic conjugates to A' and C' respectively; therefore B", A", C", are in the same straight line.

10. *If from the ends of the base of a triangle two lines be drawn to the opposite sides, so as to intersect on the perpendicular to the base, the lines joining the foot of the perpendicular to their intersections with the sides, make equal angles with the base.* For, if AA' be perpendicular to the base (see last fig.), A' is the vertex of an harmonic pencil (Art. 9, 1°) two of whose alternate legs are at right angles, and therefore (Art. 8) the angle B'A'C' is bisected, and the proposition is proved.

11. Some of the conclusions arrived at in Art. 9 may be otherwise expressed. The following definition must be first laid down:

If the opposite sides of a quadrilateral be produced to meet, the line joining their intersections may be called *the third diagonal* of the quadrilateral. Thus (see last figure), BC is the third diagonal of AC'OB'; again, C'B' is the third diagonal of ABOC; and considering C'BB'C also as a quadrilateral, AO is its third diagonal. We have then the following proposition:

"The *three diagonals* of any quadrilateral belong also to two other quadrilaterals, whose sides are segments respectively of those of the first quadrilateral; and each of them is cut in conjugate harmonic points by the other two." The figure formed by three such quadrilaterals is called by Carnot a *complete quadrilateral.*

12. Given in position one pair of opposite sides of a quadrilateral, and the point of intersection of the other pair, to find the locus of the intersection of the diagonals (that is, the diagonals as commonly understood).

Let AB and AC be the lines given in position, and A″ the given point, (see last figure); then, the intersection of the diagonals of the quadrilateral CB′C′B will evidently lie on the harmonic conjugate of A″A, with respect to AB and AC, namely AA′. This line is then the required locus.

13. *Let a righ line revolve round a given point* O, *and cut two given right lines,* VA′ *and* VB′, *in* A *and* B; *let a portion* OX *be taken on it, such that its reciprocal shall be equal to the sum of the reciprocals of* OA *and* OB (*more concisely thus:* $\frac{1}{OX} = \frac{1}{OA} + \frac{1}{OB}$);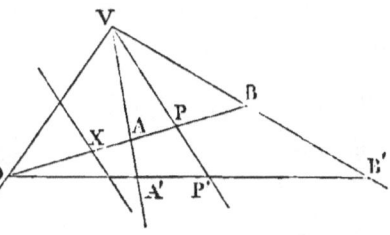
required the locus of the point X. (OX, OA, OB, are supposed to be taken in the same direction.)

Draw any line from O, and let A′ and B′ be the points where it cuts the given lines; cut A′B′ in P′ so that A′P′ : B′P′ :: OA′ : OB′ (Euclid, B. vi. Prop. 10); join OV and VP′; then *a right line* bisecting OV and parallel to VP′ *will be the locus required.* For, as OB′ is cut harmonically, OB is cut harmonically (Art. 7); therefore (Art. 5, 3°) $\frac{1}{OA} + \frac{1}{OB} = 2 \cdot \frac{1}{OP}$; but by the definition of a reciprocal (Art. 5, 3°) it is plain that the reciprocal of half a line is double the reciprocal of the whole; therefore, as the line drawn parallel to VP bisects OP, this parallel passes through the point X in all its positions. Q. E. D.

If the revolving line take such a position that B comes to the other side of V (A remaining at the same side as before), it will be found, by a similar proof, that it is cut by the same parallel in a point X, so that $\frac{1}{OX}$ equals the difference of $\frac{1}{OA}$ and $\frac{1}{OB}$. In order to reconcile this with the former result, we must recollect that OB and OA, being now measured from O in opposite directions, must have different signs (see Note on Art. 5). It frequently happens, as in this example, that some notions of algebra are required for the full comprehension of geometrical results. We shall not, however, in this work enter on that subject further than seems to be called for by questions coming under consideration.

14. Let us now suppose the revolving line to cut *any number* of given right lines in A, B, C, &c., and let OX be taken (in the same direction with OA, OB, &c.), so that $\frac{1}{OX} = \frac{1}{OA} + \frac{1}{OB} + \frac{1}{OC} +$ &c.; *the locus of X is still a right line.* For, if OX′ (see fig.) be 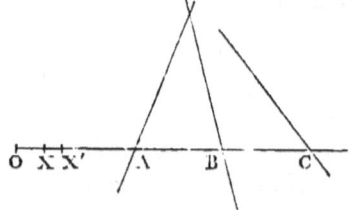 taken such that $\frac{1}{OX'} = \frac{1}{OA} + \frac{1}{OB}$, the locus of X′ is a certain right line (Art. 13), which may be considered as replacing the two right lines on which A and B are supposed to move; that is, the original condition now becomes $\frac{1}{OX} = \frac{1}{OX'} + \frac{1}{OC} +$ &c. If then the number of given lines is three, the locus of X is a right line by the last article; if four, the question is reduced to the case of three by a similar process, and so on for any number.

CHAPTER II.

ANHARMONIC RATIO AND INVOLUTION.

15. "IF four fixed right lines VA, VB, VC, VD be drawn from the same point, and any transverse right line AD be cut by them in four points A, B, C, D, the ratio of the rectangle AD · BC to the rectangle AB · CD is constant." (This ratio is called the *anharmonic ratio of the four points* A, B, C, D).

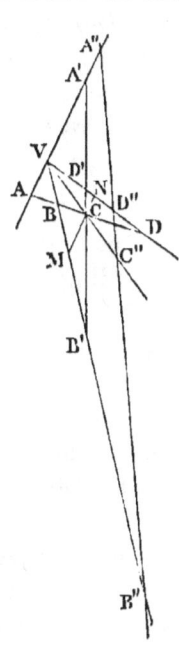

For, through C draw MN parallel to AV; then, AD · BC : AB · CD :: $\begin{Bmatrix} BC:AB \\ AD:CD \end{Bmatrix}$ :: $\begin{Bmatrix} MC:AV \\ AV:CN \end{Bmatrix}$ :: MC : CN ; but this last ratio remains constant when the point C changes its position along the line VC ; the proposition is therefore proved.

When the transversal coincides with MN, AB and AD become infinite, and are in a ratio of equality (Art. 3), and therefore AD · BC : AB · CD :: BC : CD :: MC : CN, as before.

If the transverse line be drawn through C, so as to intersect AV produced (see fig.) through V, we shall still find that A'D' · CB' : A'B' · CD' :: MC : CN ; and if C moves along VC into any other position C", we shall have for any line such as A"B", A"D" · B"C" : A"B" · C"D" :: AD · BC : AB · CD.

From the last remark it immediately follows that, *if the four given lines be all produced through* V, *the ratio of the rectangles under the segments of any transversal cutting them in any way will always be the same, the rectangles being written as above, and the same letter* A, B, C, *or* D, *being used to represent a point anywhere along the whole extent of one and the same right line.* In other words, the anharmonic ratio of the four points on the transversal will remain constant.

The right lines meeting in a point are said to form a pencil; and the constant ratio above-mentioned is called *the anharmonic ratio of the pencil.* This may be expressed by the notation V · ABCD, V being the vertex of the pencil.

It is evident that, *if two pencils cross the same right line in the same four points, the anharmonic ratios of the pencils are equal,* both being equal to that of the four points.

It is also evident, that *if the angles contained by corresponding legs of two pencils be equal, the pencils have the same anharmonic ratio.*

The great importance of the anharmonic properties of a pencil was first indicated by Chasles in the Notes to his "Aperçu Historique." The principle itself is given in the Math. Col. of Pappus, B. vii. Prop. 129.

16. There are two other constant ratios deducible from the preceding.

For, the sum of the rectangles AD · BC and AB · CD (see last fig.) = AB · BC + CD · BC + AB · CD = AC · BC + AC · CD = AC · BD. We have also AD · BC : AD · BC + AB · CD :: MC : MC + CN. Hence AD · BC : AC · BD :: MC : MN; and (in like manner) AB · CD : AC · BD :: CN : MN; both constant ratios.

It follows from this, that *when a given pencil is cut by a variable transversal, the ratio* AD · BC : AB · CD *is still a constant, even when the letters* A, B, C, D, *have been interchanged among the legs in any way.* The value of the constant will not, of course, be the same during all the interchanges.

It is evident that, when MN is bisected, the pencil is harmonic. But the subject of harmonic pencils is in itself so important that we have treated it separately.*

---

* *The anharmonic ratio of a pencil admits of a remarkable trigonometrical expression,* which is easily deduced from the geometrical.

For (see fig. of Art. 15), $\dfrac{MC}{CN} = \dfrac{MC}{CV} \cdot \dfrac{CV}{CN} = \dfrac{\sin BVC}{\sin AVB} \cdot \dfrac{\sin AVD}{\sin CVD}$; therefore $\dfrac{AD \cdot BC}{AB \cdot CD} = \dfrac{\sin AVD \cdot \sin BVC}{\sin AVB \cdot \sin CVD}$; the required result.

If the figure represent an harmonic pencil, we shall have $\sin AVD \cdot \sin BVC = \sin AVB \cdot \sin CVD$.

17. *Given in position three legs of a pencil, its anharmonic ratio, and the relative position of the fourth leg, (that is, between which successive pair of the given legs it lies), its actual position is determined.*

For, it is easy to see that in all cases the line VA (see fig. Art. 15) may be taken to represent one of the given legs, and that, if we draw any right line, MN, parallel to it, two of the three points, M, C, N, will be given, as well as the ratio, MC : NC, by which the third point is determined, and therefore the fourth leg also.

EXAMPLES ON ANHARMONIC RATIO.

18. We shall now exemplify the foregoing principles by the following proposition.

"Given in magnitude and position the bases AE, BD of two triangles, AFE, BCD, whose vertices, F, C, move each along the base of the other triangle, the line joining P the intersection of the sides AF, DC, to Q, the intersection of EF, BC, constantly passes through a fixed point O, the intersection of the lines AB and ED."

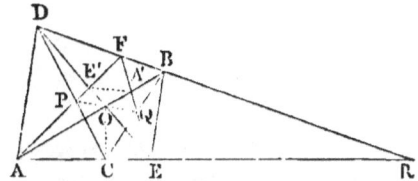

For, draw AD, EB, OP, OQ, OC, and let DB and AE meet at R. Then the anharmonic ratio of the pencil OC, OE, OQ, OB (which we shall express more concisely by O · CEQB), is the same as that of the pencil EC, EO, EQ, EB (Art. 15), that is, O · CEQB = E · COQB; but this latter anharmonic ratio equals A · RDFB (Art. 15), which again equals O · CDPA; therefore O · CEQB = O · CDPA; and therefore (Art. 17) P, O, Q must be in one right line. *Q. E. D.*

If we call P′ the intersection of AF with BC, and Q′ that

In a similar manner we should find (for any pencil) $\dfrac{AD \cdot BC}{AC \cdot BD}$ =(see same figure)
$\dfrac{\sin AVD \cdot \sin BVC}{\sin AVC \cdot \sin BVD}$, and $\dfrac{AB \cdot CD}{AC \cdot BD} = \dfrac{\sin AVB \cdot \sin CVD}{\sin AVC \cdot \sin BVD}$.

Since AD · BC + AB · CD = AC · BD, the last two equations give, by addition, sin AVD · sin BVC + sin AVB · sin CVD = sin AVC · sin BVD.

This equation expresses the relation existing between the sines of the six angles, made by four right lines which meet in a point.

of EF with DC, we shall find, in a similar manner, that the line joining P′, Q′ constantly passes through a fixed point, namely, the intersection of AD with BE.

The student is recommended to draw separate figures representing the varieties in the construction given above, as, for instance, when the point F lies on the production of BD, or when the lines AE, BD are given, cutting one another at a point between A and E. By using the same letters as before to express the corresponding points in the new figure, there will be no difficulty in adapting the proof above given to any other case. It is only by a practice of this kind that the full extent and meaning of geometrical propositions can be understood.

19. The preceding theorem leads to several remarkable results:

1°. If CF be drawn, we have the following enunciation:

Let a given quadrilateral be divided into any two others; the line joining the intersection of the diagonals of one of them, with the intersection of those of the other, passes through the intersection of the diagonals of the given quadrilateral.

2°. Again, let all the lines on the figure be considered fixed, except BC, DC, PQ, OC; then we find that,

If the angles of a variable triangle PCQ move on three given right lines, FA, AE, EF, and if two sides CP, CQ pass respectively through two given points D, B, in directum with the intersection F of the two given lines on which the opposite angles move, the third side PQ also passes through a fixed point O.

If, as before (Art. 18), we call P′ the intersection of AF and BC, and Q′ that of EF and DC, we shall have another variable triangle P′CQ′, to which the preceding enunciation still applies, by changing P and Q into P′ and Q′, interchanging the points D and B, and substituting the intersection of AD and BE (which we shall call O′) in place of O. The whole series of variable triangles, satisfying the given conditions, consists, therefore, of two distinct systems, corresponding to the two distinct points O and O′.

3°. Conversely, *if the sides of a variable triangle PCQ pass through three fixed points D, B, O, and if two of the angles P, Q move on two given lines FA, FE, whose intersection is in directum with the two points through which the opposite sides pass, the locus of the third angle* C *is a right line* AE. The establishment of this we leave to the reader.

Again, if we conceive DQ and BP to be drawn, and to intersect at a point C′, we shall have another variable triangle QC′P, for which the preceding enunciation still holds good, by changing C into C′, and substituting in place of AE the line A′E′ (see last figure); and, therefore, the *complete* locus of the third angle of a triangle, moving under the prescribed conditions, consists, not of one right line, but of two.

4°. If we take the six points A, F, E, D, C, B, and by (joining each successive pair AF, FE, ED, DC, CB, BA) consider them as the six vertices of a hexagon, the theorem in Art. 18 becomes a case of Pascal's theorem concerning a hexagon inscribed in a conic section; that is, *if the six vertices of a hexagon lie on two right lines, the intersections of the opposite sides,* (1 *and* 4, 2 *and* 5, 3 *and* 6) *lie in a right line.*

20. *If two pencils have one leg common, and the same anharmonic ratio, their vertices being different, the intersections of the corresponding legs lie on a right line.*

For, join two of the intersections, and produce the joining line to meet the common leg; this joining line must meet the fourth leg of each pencil in the same point (by what has been said in Art. 17); and the proposition is proved.

21. The principle just established is often useful: we shall take the following theorem as an example:

If the sides of a variable triangle pass through three given points in a right line, and if two angles move on given right lines, the third angle will always lie on one of two definite right lines, passing through the intersection of the two given lines.

Let Q, R, S be the given points, and VA, VA′ the given lines. Take three positions of the moving triangle, and let BOB′ be one of them.

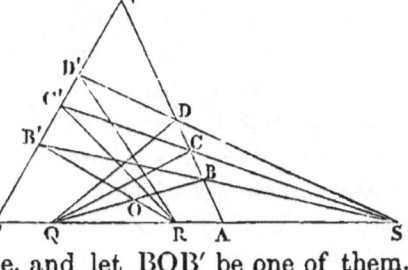

Then the anharmonic ratio Q · ABCD = R · A'B'C'D', each of them being equal to S · ABCD (Art. 15); therefore (Art. 20) the three positions of the point O lie in a right line. Now, let two of the positions of O be considered as fixed, and the third as variable; then the last evidently moves on a right line.

Again, if we conceive B to be joined to R, and B' to Q, and call O' the intersection of the joining lines, we shall have another variable triangle B'OB satisfying the prescribed conditions; and it will appear, as before, that O' moves on a definite right line. It is evident, in both cases, that when the points B, B' coincide with V, the point O or O' will also coincide with V; and therefore the two right lines on which O and O' move must pass through V.

If we had taken only two positions of the moving triangle, the proof might have been completed by joining V to Q, R, S. In this case the point V represents an evanescent state of the triangle.

22. Conversely, *if the three angles of a variable triangle move on three given right lines, which meet in a point, and if two sides pass respectively through two given points, the third side will also pass through one of two definite points in directum with the given points.* The student will find no difficulty in establishing this result.

We shall return in the next chapter to the relation existing between this proposition and that given in the last Article.

COPOLAR TRIANGLES.

23. As another example of the application of the principle contained in Art. 20, we shall take the following proposition, due to Desargues:

If two triangles, ABC, A'B'C' be such that the lines joining corresponding vertices AA', BB', CC', meet in a point O, the intersections of corresponding sides (of BC and B'C', &c.), lie on one right line; and conversely.

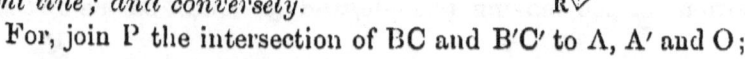

For, join P the intersection of BC and B'C' to A, A' and O;

then the anharmonic ratio $A \cdot PCA'B = A' \cdot PC'AB'$, each of these ratios being equal to $O \cdot PCA'B$ (Art. 15); and therefore (Art. 20) the three intersections P, Q, R are *in directum*. The first part of the proposition is therefore proved.

To prove the converse, let CC' and BB' meet in O. Now, as P, Q, R are given in a right line, it follows, by applying the part already proved to the triangles QCC', RBB', that the points O, A, A' are in one right line; that is to say, "if the intersections of corresponding sides of the two triangles ABC, A'B'C' be *in directum*, the lines AA', BB', CC', joining corresponding vertices, meet in one point."

24. With respect to Desargues's theorems given in the last Article, it is remarkable that the conclusions are equally correct when the two triangles are situated *in different planes*. This may be proved *a priori* as follows: In the first part, it is granted that CC' and BB' (see last figure) lie in the same plane; therefore, so also do BC and B'C'; the latter lines, therefore, meet one another on the line of intersection of the planes of the triangles. The same reasoning applies to the other pairs of corresponding sides, and, consequently, the first theorem is proved. The second follows from the first, exactly as before.

As these results are independent of the magnitude of the angle made by the planes, they will hold good when the angle vanishes; and thus we have an independent proof of the proposition contained in Art. 23.

The proposition referred to is sometimes stated in the following manner: "Two copolar triangles are co-axial, and *vice versâ*."

ANHARMONIC PROPERTIES OF THE CIRCLE.

25. The anharmonic properties of the circle depend on the following principle:—

If four fixed points, A, B, C, D, be taken in the circumference of a circle, and lines be drawn from them to a variable fifth point O, also in the circumference, the anharmonic ratio of the pencil so formed with O, as vertex, is constant. (This is called the anharmonic ratio of the four points on the circle.)

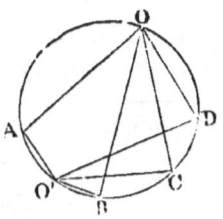

For, while O remains between A and D, all the angles at O

remain unchanged in magnitude, and the proposition is evident (Art. 15). Now, let us suppose O to pass beyond A into the position O′. In this case the angles made by O′ with O′D, O′C, O′B are equal respectively to those made by OD, OC, OB, with the production of AO through O (Euclid, B. iii. Prop. 22); therefore O′ . ABCD = O . ABCD (Art. 15), and the truth of the proposition is manifest.

26. *To express the anharmonic ratio of four points on a circle* (see Art. 25) *by means of the chords joining the given points.*

Let O be so taken that AB′=AB (see fig.); Draw B′Q parallel to CD, and join A to C and Q. Then, since the angle B′QC′ is equal to the angle C′CD, equal to the angle OAB′, AOQB′ is capable of being circumscribed by a circle (Euclid, B. iii. Prop. 22); and therefore the angle AQB′ = AOB′ = ACB; also, the angle AB′Q = ABC (Euclid, B. iii. Prop. 22); therefore (Euclid, B. i. Prop. 26) B′Q = BC, and therefore B′C′ : C′D :: BC : CD. Again, we have AD : AB′ :: AD : AB. Therefore, compounding the ratios, AD . B′C′ : AB′ . C′D :: AD . BC : AB . CD, which last ratio gives the required result.*

27. *If a hexagon be inscribed in a circle, the intersections of the opposite sides are three points in one right line.*

This is Pascal's theorem applied to the circle.

Let P, Q, R be the intersections of the opposite sides of the inscribed hexagon AO′BCDO; join QC, OB, OC, O′C, O′D; then Q . BCDR = O . BCDR = O . BCDA = (Art. 25) O′ . BCDA = O′ . BCDP = Q . BCDP; and therefore (Art. 17) P, Q, R are in one right line. (It will be convenient to call this line a Pascal's line.)

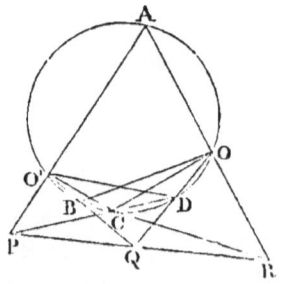

---

* If R be the radius of the circle, circumscribed to a triangle, it is easy to see that any side equals 2R multiplied by the sine of the opposite angle; therefore, the result of this Article follows from the trigonometrical mode of expressing the anharmonic ratio of a pencil given in the Note on Art. 16.

M. Chasles has applied similar principles to prove Pascal's theorem in its general form. (Aperçu Historique, p. 336.)

28. If the points A and O become coincident, AO becomes a tangent, and we have the following theorem:

"If a pentagon be inscribed in a circle, and a tangent be drawn at one angle to meet the opposite side, and if the sides containing the angle be produced to meet the remaining pair of sides, the three points of intersection, so found, are in one right line."

If B and C also become coincident, we find that

"If a quadrilateral be inscribed in a circle, tangents at the extremities of either diagonal, intersect on the *third diagonal*." (See Art. 11.)

When three alternate sides of the hexagon vanish, the theorem becomes the following:

"If a triangle is inscribed in a circle, tangents at the angles intersect the opposite sides in three points in one right line."

29. There is another form of the theorem given in Art. 27, which leads to some remarkable results.

Let the six points ABCDEF (see fig.) be joined consecutively, A to B, B to C, and so on, ending with the line FA; the successive joining lines may be considered as sides (1, 2, 3, 4, 5, 6) of a hexagon, to which Pascal's theorem will still apply. For the satisfaction of those not yet accustomed to this way of considering polygons, we shall give a distinct demonstration.

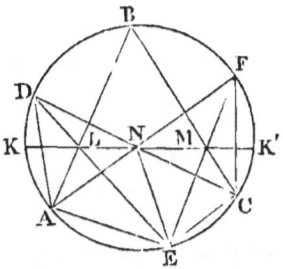

Join CF, CE, AD, AE, LN, MN. Then, N · FMCE = C · FBDE (Art. 15), = A · FBDE (Art. 25), = N · ALDE; therefore (Art. 17), LN, NM, are *in directum*; that is, the intersections L, M, N of opposite sides (1 and 4, 2 and 5, 3 and 6) are *in directum*. Q. E. D. The following consequences are evident:—

It also follows from the Note referred to (Euclid, B. iii. Prop. 20), that the anharmonic ratio of the four points A, B, C, D, on the circle, equals $\dfrac{\sin \frac{1}{2} AD \cdot \sin \frac{1}{2} BC}{\sin \frac{1}{2} AD \cdot \sin \frac{1}{2} CD}$.

If DF and AC be drawn, we shall have two triangles, ABC, DEF, inscribed in a circle; let the bases AC, DF be given in magnitude and position; we find, then, that the line joining L and M, the respective intersections of the sides, constantly passes through a fixed point.

In like manner:—If AB cuts FE in L', and BC cuts DE in M', the line joining L' and M', will constantly pass through the intersection of AD and FC.

Again, let all the lines on the figure be fixed, except DE FE, LM, NE; then, we perceive that a variable triangle MEL, whose sides pass through three given points F, D, N, and having two of its angles L and M on two given right lines, will have for the locus of its third angle E a circle, provided that the data be such, that a certain pentagon ADBFC will admit of a circle being circumscribed to it.

And conversely, it appears that, under certain circumstances, a triangle, two of whose angles move on given right lines, while the third moves on a given circle, and two of whose sides pass respectively through two given points, will have its third side also constantly passing through a fixed point.

PROBLEMS RELATING TO ANHARMONIC RATIO.

30. The preceding construction enables us to solve the following problem:

Given six points on a circle, to find a seventh (also on the circle), such that the anharmonic ratio of it, with three of the given points taken in a definite order, shall be equal to that of it, with the remaining three, taken also in a definite order.

For example, let it be required (see last figure) to find K, so that the anharmonic ratio of K, A, E, C, shall be equal to that of K, D, B, F.

Construct a hexagon with the six given points, in such a manner that the first set of points, A, E, C, shall be opposite vertices to the second set, D, B, F, respectively. (In the present case, the hexagon ABCDEF is the proper one, the first vertex, A, being opposite to the fourth, D, the second, B, to the fifth, E, and the third, C, to the sixth, F.) The Pascal's line (see Art.

27) belonging to this hexagon, will cut the circle in two points, K, K', either of which will answer the question.

For, if we suppose KD and KA to be drawn, we shall have Art. 15) $D \cdot KAEC = A \cdot KDBF$; therefore (Art. 25), the point K answers the question, and in a similar manner it may be proved for the point K'.

31. If, in the enunciation of the foregoing problem, we substitute *a right line* in place of a circle, we shall have another important problem, which is easily reduced to the former.

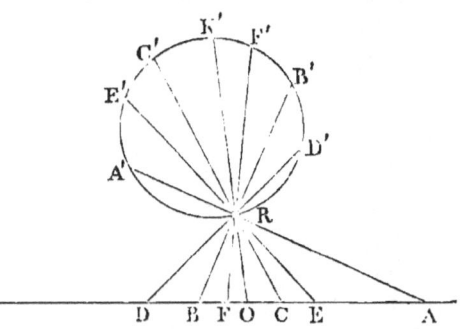

Let A, E, C, and D, B, F, be the two sets of points, now *given in a right line;* take any circle, and any point R on it, and join R to the six given points. We shall have, thus, six points, A', E', C', D', B', F', on the circle. Find a seventh point K or K' on the circle, according to the conditions of the last problem, and join it to R; the joining line will cut the given line in a point O, answering the present question. This appears from Arts. 25 and 15. For, $R \cdot AECO = R \cdot A'E'C'K' = R \cdot D'B'F'K' = R \cdot DBFO$.

32. We can now give the solution of a very remarkable problem, viz.:—*To inscribe, in a given polygon, another of the same number of sides, so that each side may pass through a given point.*

That the figure may not be too complicated, we shall take the case of a triangle, but the argument will equally apply to any polygon.

Let P, Q, R (see figure on next page) be the three points through which the sides of the inscribed triangle KLM are to pass. Take on one of the sides of the given triangle three points M_1, M_2, M_3, as positions (on trial) of one angle of the required triangle, and construct three triangles whose sides shall pass through the given points, and whose second set of angles, L_1, L_2, L_3,

shall lie on a second side of the given triangle. The sides containing the third set of angles will cut the third side of the triangle in six points, A, D, E, B, C, F. Now (Arts. 15 and 16), $P \cdot KAEC =$
$P \cdot LL_1L_2L_3 =$
$Q \cdot LL_1L_2L_3 =$
$Q \cdot MM_1M_2M_3 =$
$R \cdot MM_1M_2M_3 =$
$R \cdot KDBF$;
therefore the anharmonic ratio of K, A, E, C, equals that of K, D, B, F; the finding of the point K is therefore reduced to the problem given in Art. 31, and the solution is complete.*

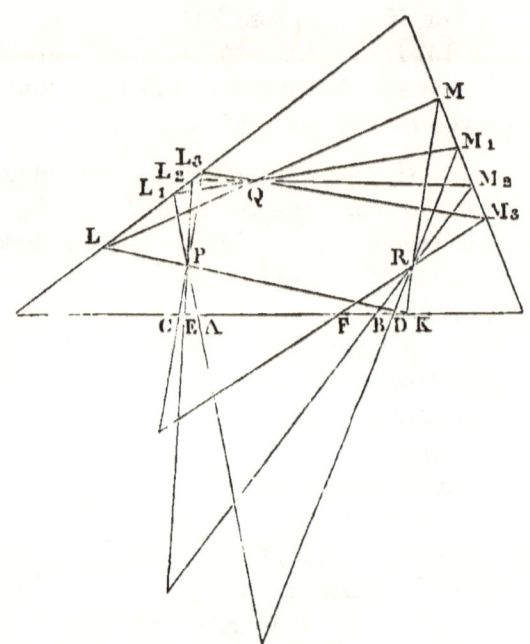

The applications of anharmonic properties given in the last three articles are due to Mr. Townsend.

HEXAGON INSCRIBED IN A CIRCLE.

33. We shall now return to the consideration of the hexagon inscribed in a circle. (See Arts. 27, 28, 29.)

---

* In the solution above given, the successive order of the angles and sides of the polygon to be inscribed, with respect to the sides of the given polygon and the given points, is supposed to be assigned. On this supposition the number of solutions is, in general, *two*, since (Art. 31) the point K admits of two positions on the side of the given polygon. When the given points, P, Q, R, &c. lie on a right line, it is easily proved that one of the positions of K coincides with the intersection of that line with the side of the polygon; and then the only *effective* solution is that determined by the other position of K. There is also but one effective solution when the given polygon vanishes into a point, that is, when the lines on which the angles of the required polygon are to lie meet in a point. In this case it will be seen that the point itself is one of the positions of K on any of the given lines, and that the other position must be taken in order to obtain an *effective* solution.

When the successive order of the angles and sides of the required polygon is not

1°. Let us find how many Pascal's lines there are for inscribed hexagons having the same vertices, A, B, C, D, E, F.

When three points are given, we can form, by joining them, but one triangle. When a fourth point is given, we can make three quadrilaterals by joining it to the extremities of each side of the triangle taken in turn. In a similar way, it appears that we can make, with five points, $3 \cdot 4$ or 12 pentagons; and with six points, $3 \cdot 4 \cdot 5$ or 60 hexagons. Each inscribed hexagon has its own Pascal's line; the number required is therefore 60.

2°. Let us next find the number of distinct points of intersection of the opposite sides.

In the first place we shall prove that through every such point four Pascal's lines can be drawn. For, considering the four hexagons ABCDEF, ABFDEC, ABCEDF, ABFEDC, the intersection of the first and fourth sides, AB, DE, is the same for all; and in the same way any other pair of opposite sides will always belong to four distinct hexagons.

This being understood, it follows that the total number of distinct points of intersection of opposite sides equals $\dfrac{60 \cdot 3}{4} = 45$.

3°. We shall conclude this subject with one of Steiner's theorems (see Salmon's Conic Sections, p. 322):—

"The sixty Pascal's lines consist of twenty sets of three, each set passing through a point."

In order to prove this, let us consider (see figure) the two triangles PQR, P'Q'R'. We see that their corresponding sides intersect in three points, L, N, M, in one right line (Art. 29), and therefore (Art. 23) the lines PP', QQ', RR', joining corresponding angles, meet in a point. But PP' is the Pascal's

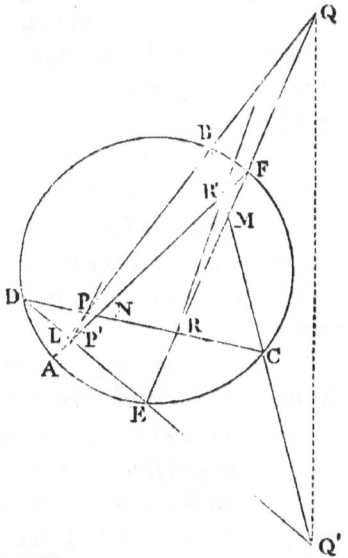

assigned, the total number of solutions equals (in general) $2^2 \cdot 3^2 \ldots (m-1)^2$, m being the number of the given points. (Poncelet, Traité des Propriétés Projectives, p. 348.) This expression must be divided by two in the two particular cases before mentioned.

line of the hexagon ABEDCF; QQ' is that of the hexagon ADEFCB; and RR' belongs to the hexagon AFEBCD. In the same manner the proposition can be proved for another set of three, and so on.

All the sets may be formed by the following rule:

"Take three of the vertices as the first, third, and fifth (always beginning with A), and put down the other three as second, fourth, and sixth, taking them in their order in the alphabet, and making each of the three, in turn, the second." Thus, ACBEDF, AEBFDC, AFBCDE, represent a single set.

INVOLUTION.

34. We shall now explain the principles of Geometrical Involution.

LEMMA 3.—Given two pairs of points, A, A', and B, B', in a right line, a fifth point, O, can be found in it, such that $OA \cdot OA' = OB \cdot OB'$.

Draw any two lines through A and B parallel to one another, and two more through A' and B'; the line joining the intersections M, N (see figures), will cut the given line in the required point.

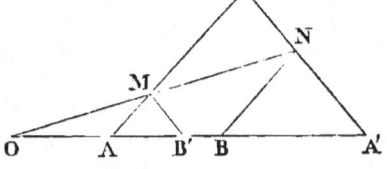

For, OA : OB :: OM : ON :: OB' : OA'; therefore

$OA \cdot OA' = OB \cdot OB'$.

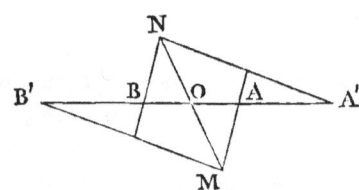

We shall now lay down the following fundamental proposition:

Given two pairs of points, A, A' and B, B', and also a fifth point, C, in a right line, a sixth point C' can be found in it, such that the anharmonic ratio of any four whatever of the six points shall be equal to that of their four conjugates. (The points forming any pair A, A', or B, B', or C, C', are said to be *conjugate*.)

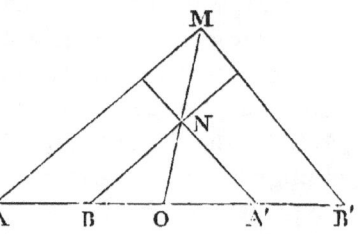

This proposition may be divided into two cases, according to the relative positions of the four given points A, A', B, B'.

First case.
—When both points of one pair lie between those of the other pair, or else when no point of either pair is so situated.

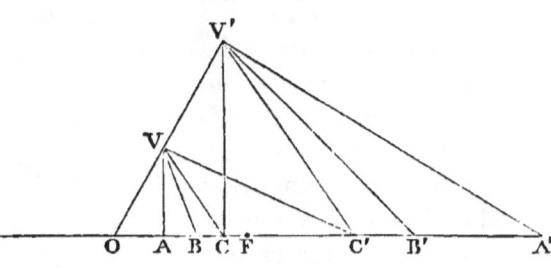

(The figure represents B and B' between A and A'.)

Let O be found (by the foregoing Lemma), so that $OA \cdot OA' = OB \cdot OB'$, and let another point, C', be taken (on the *same* side of O as C is), such that $OC \cdot OC' = OA \cdot OA'$; then C' is the point in question.

For, draw any line from O, and take in it two points, V, V' (on the *same* side of O), such that $OV \cdot OV' = OA \cdot OA'$; and join VA, VB, VC, V'A', V'B', V'C', VC', V'C. It is evident, from the construction, that the quadrilaterals AA'V'V, BB'V'V, CC'V'V, are inscribable in circles (Euclid, B. iii. Prop. 36); therefore, the angle OVA equals the angle OA'V' (Euclid, B. iii. Prop. 22); also, the angle OVB = OB'V'; and therefore the angle AVB = A'V'B'. In like manner the angle BVC = B'V'C', and CVC' = CV'C'; therefore (Art. 15) $V \cdot ABCC' = V' \cdot A'B'C'C$. In the same way, by joining V and V' to the other points on the line, the anharmonic ratio of any other set of four points may be proved equal to that of their four conjugates. Q. E. D.

When no point of either pair lies between the points of the other pair, B and B' will be at a different side of O relatively to A and A'; the angle AVB will be the supplement of A'V'B'; and the angle BVC the supplement of B'V'C'. As, however, the angle between two right lines, considered as indefinitely prolonged, may be a certain angle, or its supplement, it follows, from Art. 15, that the result will be the same as before.

Should the given point C and the point A not be at the same side of O (as we have hitherto supposed) there will be no difference in the result.

The student is recommended to examine separately the circumstances indicated in the preceding remarks.

Second Case.—When a single point of a pair (suppose A′) lies between the points of the other pair B, B′. (See figure.)

In this case, the proof is completely analogous to that of the former, observing that C′ and C, as well as V′ and V, are now to be taken on *different* sides of O.

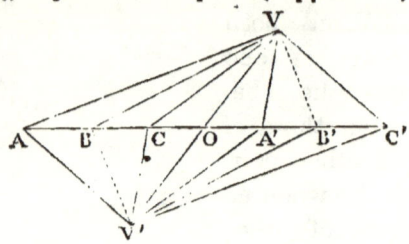

35. *If three pairs of points in a right line be such that the anharmonic ratio of four of the points equals that of their four conjugates, and if the relative order of the two sets of points be also the same, the anharmonic ratio of any other set of four is equal to that of their four conjugates.* (Six points in a right line, having this relation, are said to be in *Involution*.)

For, five of the six points can be taken as the five points in the preceding proposition, such that, by Art. 17, the sixth point must coincide with the sixth point there determined; and the proposition then becomes manifest.

From what has been said it follows, that *when three pairs of points A and A′, B and B′, C and C′, are in involution, there always exists another point O in the same right line with them, such that* $OA \cdot OA' = OB \cdot OB' = OC \cdot OC'$, *the points of every pair being, moreover, situated both on the same side of this point O, or else on opposite sides of it.*

And conversely, *if three pairs of points in a right line satisfy the foregoing conditions, they are in involution.*

36. Let us now consider the points C and C′ as variable under the conditions just referred to, and we shall have *an indefinite number of pairs of points, each of which is in involution with the given pairs A, A′ and, B, B′. Any three pairs of the entire system will also be in involution.*

There are, besides, some other remarkable properties of such a system, which we subjoin.

1°. The point O (which is called the *centre* of the system) is the conjugate of a point at infinity.

This is evident from the condition, $OC \cdot OC' =$ constant. For, if OC' becomes infinite, OC must vanish.

2°. A point which coincides with its conjugate is called by Chasles a *double point*, and by Mr. Davies a *focus* of the system. Such a point can only exist in the first of the two cases mentioned in Art. 34. In that case, there are two such points, F, F', at equal distances on different sides of the centre, the distance OF being given by the equation $OF^2 = OA \cdot OA'$. (See the figure of the first case in Art. 34.)

It follows from this (Art. 3) that *any pair of conjugate points*, in the present sense of the term, *are harmonic conjugates in respect to the two foci*. Hence (Art. 3), if one focus bisect the distance between one pair of conjugates, the other focus is at an infinite distance, and therefore the distance between any other pair is also bisected.

3°. When the foci are *real*, a circle described on FF', as diameter, is the locus of a point, at which the angle subtended by any two of the system of points is *equal or supplemental* to that subtended by their conjugates, according as the two points lie at the *same or different* sides of the centre.

This is evident by supposing V and V' to coincide in the first case of Art. 34.

The method here followed in explaining the properties of involution is a modification of that given by Mr. Townsend (Salmon's Conic Sections, Art. 317). In the construction given in the work referred to, the points V, V' coincide, which consequently restricts its application to the first case of Art. 34. The analogous conclusions respecting the second case, may be deduced by the aid of the principle of continuity; but as that principle cannot be fully apprehended by those not yet acquainted with the application of algebra to geometry, we have thought it advisable to give the demonstration of Art. 35 in a form directly applicable to all cases.

EXAMPLES ON INVOLUTION.

37. We shall conclude this Chapter with some examples on the foregoing principles.

1°. "Lines drawn from any point in the plane of a quadrilateral to the six points of intersection of the sides form *a pencil in involution*," that is, a pencil of six legs cutting any transversal in involution. (It is evident, from Arts. 15 and 35, that if one transversal be so divided, the same will be true of any other.)

Let ABA'B' be the quadrilateral, C, C' the intersections of the opposite sides, and V the given point. Let VA, VA' be taken as conjugate legs; also, VB, VB' and VC, VC'. Join BB'; then $V \cdot AB'BC' = B \cdot AB'VA'$ (Art. 15), $= V \cdot CB'BA' = V \cdot A'BB'C$; therefore, if we conceive a transversal to cut 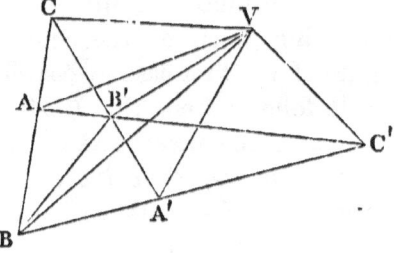 the pencil of six legs, having V as vertex, it will cut in involution (Art 35), and therefore the pencil itself is in involution. Q. E. D.

2°. *A right line, AA', cutting the sides and diagonals of any quadrilateral, RVSV', is cut in involution.*

Let A, A' be taken as conjugate points; also (see fig.), B, B' and C, C'. Join VC' and V'C'. Then, $V \cdot ABCC' = V \cdot SRCC' = V' \cdot SRCC' = V' \cdot B'A'CC' = V' \cdot A'B'C'C$; therefore (Art. 35), AA' is cut in involution.

It appears from Art. 36, 2°, that if the transversal be drawn 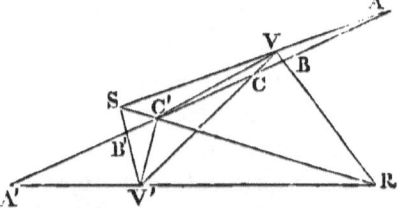 through the intersection of the diagonals, that point will be one of the foci of the system determined by the two pairs, A, A' and B, B'. If, in addition, AA' is bisected by the intersection of the diagonals, so also will BB'. In this case the other focus, and the centre of the system, will be at an infinite distance.

3°. *A right line cutting a circle, and the sides of a quadrilateral, PQRS, inscribed in the circle, is cut in involution.*

Join QB, QB', SB, SB', (see figure on next page); then, $Q \cdot ABCB' = Q \cdot PBRB' = $ (Art. 25) $S \cdot PBRB' = S \cdot C'BA'B' = S \cdot A'B'C'B$; therefore (Art. 35) AA' is cut in involution.

If the diagonals of the quadrilateral be drawn, and the points

where they cut the transversal be called D and D', we shall have
(by Prop. 2°) the two pairs of points, A,
A' and C, C', in involution with D, D';
but they are also in involution with B, B';
therefore (Art. 36) any three of the four
pairs are in involution.

4°. *If three chords of a circle intersect
in a point, lines drawn from any point
in the circumference to the extremities of
the chords, form a pencil in involution.*
(See Ex. 1°.)

Let AA', BB', CC', be the three chords.
Join AC, AC', AB', CB'. Then, by Art. 15,
A · CA'B'C' = B' · CBAC' = B' · C'ABC;
therefore (Art. 25), if V be any point on
circumference, V · CA'B'C' = V · C'ABC,
and the proposition becomes evident by
Art. 35.

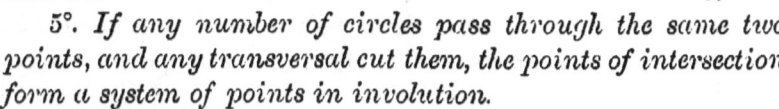

5°. *If any number of circles pass through the same two
points, and any transversal cut them, the points of intersection
form a system of points in involution.*

Let the circles intersect in P, Q (see fig), and let the line
joining P, Q cut the transver-
sal in O. Then (Euclid, B. iii.
Props. 35, 36), OA · OA' = OP ·
OQ = OB · OB' therefore OA ·
OA', OB · OB', OC · OC', &c.,
are all equal, and therefore, by
Arts. 35 and 36, the system of
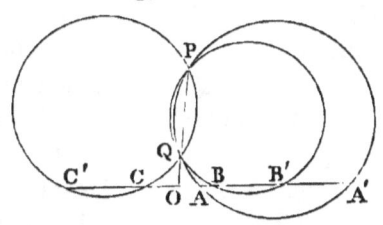
points AA', BB', CC', &c., is in involution, O being the centre
of the system.

It is evident, (Euclid, B. iii. Prop. 36), that if two circles be
described through P and Q, to touch the transversal, the points
of contact will be the foci of the system. These will, of course,
be *imaginary* (Art. 36), when the point O lies between P and Q.

If the transversal be parallel to PQ, one focus and the centre
of the system are at an infinite distance.

CHAPTER III.

POLES AND POLARS IN RELATION TO A CIRCLE.

38. GIVEN a point O, and a circle whose centre is C (see next figure); let CO be joined, and, on the joining line, a portion, CO', be taken (in the same direction from C as O is), such that CO · CO' is equal to the square of the radius of the circle; a perpendicular to the line CO', at the point O', is called the *polar* of the point O, in relation to the given circle, and the point O is called the *pole* of the perpendicular.

Hence, *when the pole is within the circle, the polar is without it; when the pole is on the circle, its polar is the tangent drawn at the pole; and, when the pole is without the circle, its polar coincides with the chord of contact of tangents to the circle drawn from the pole* (Euclid, B. vi. Prop. 8).

39. *Any right line through the pole is cut harmonically by the circle and the polar.* (The right line is supposed to meet the circle.)

Let us first consider the case when the pole is *within* the circle. Let PR be the line (see fig.). Join CP, CQ, O'P, O'Q; then, since CO · CO' = CQ², we have CO : CQ :: CQ : CO', and, therefore (Euclid, B. vi. Prop. 6), the angle CQO is equal to the angle CO'Q; in the same way the angle CPO is equal to the angle CO'P, and therefore the angles CO'Q and CO'P are equal, and also their complements; therefore, O'O and O'R are the internal and external bisectors of the angle PO'Q, and therefore PR is cut harmonically (Lardner's Euclid, B. vi. Prop. 3).

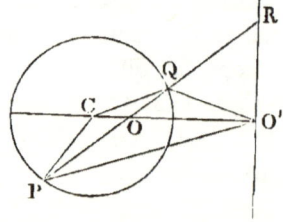

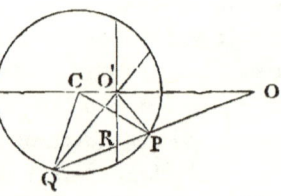

Let us now suppose the pole to be without the circle. Let OQ be the line. As before, we find the angle CQO = CO'Q, and the angle CPO = CO′P; but the angles CQO and CPO are supplements; therefore CO′Q and CO′P are supplements; therefore, the angles, OO′P and CO′Q, are equal, and also their complements; therefore, O′O and O′R are the external and internal bisectors of the angle PO′Q, and the line OQ is cut harmonically.

40. From the preceding Proposition, very important consequences may be deduced.

1°. *If any number of points, on a right line, be taken as poles, their polars, with respect to a given circle, pass through one point, namely, the pole of the right line.*

For, taking O′R as the right line, and R as one of the points (see the preceding figures), it follows, from the last Proposition, that the polar of R must pass through O, since the line through R and O is cut harmonically.

From this it follows (Art. 38), that *if pairs of tangents be drawn to a circle from all the points of a given right line, the chords of contact pass through one point.*

Also, *the line joining any two points has, for its pole, the intersection of the polars of the points.*

2°. *If any number of right lines pass through a point, the locus of their poles, with respect to a given circle, is a right line, namely, the polar of the point.*

For, let O (see the preceding figures) be the point; conceive any line to be drawn through O, and its pole to be joined to that point; then, as the joining line is cut harmonically, the pole of the line drawn through O, must lie on O′R.

Hence, it appears, that "if any number of lines be drawn through a given point, so as to cut a given circle, and tangents be drawn at the points of intersection of each line with the circle, the locus of their intersection is a right line, namely, the polar of the given point."

The reader may, perhaps, imagine that Props. 1° and 2° are not to be relied on in all cases, as the proofs above given will sometimes require us to consider a line cut harmonically with one pair of conjugates *imaginary*, (namely, when the points R and O are both without the circle). The validity of the proofs

F

depends on the principle of continuity, which we shall endeavour to explain in a future Chapter. Another proof of Props. 1° and 2° is given below.*

3°. *If through a given point, A, any two right lines be drawn, cutting a given circle in the points B, C, and B′, C′, the lines joining those points (whether directly or transversely) will intersect on the polar of the point A.*

Let BB′, CC′, intersect at O (see fig.). Join AO; then, taking OB and OC as two conjugate legs of an harmonic pencil, and OA as a third leg, the fourth leg OD is determined (Art. 7), which, as it cuts AC and AC′ harmonically, must be the polar of A. The point O lies, therefore, on the polar of A; and by joining AO′, it follows, in the same way, that O′ also lies on the polar of A.

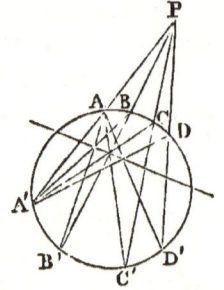

The figure represents A without the circle, but the proof applies in every case.

4°. *If four right lines be drawn from the same point, P, so as to cut a circle, the anharmonic ratio of any four of the points of intersection (see Art. 25), as A, B, C, D, is the same as that of the remaining four, A′, B′, C′, D′.*

Join A to B′, C′, D′ (see fig.), and A′ to B, C, D. Then, the intersections of the joining lines respectively (AB′ with A′B, AC′ with A′C, and AD′ with A′D), lie on the polar of P (Prop. 3°); therefore A . A′B′C′D′

---

* 1°. Let O′S be a given right line, of which O is the pole, and let S be any point on it. Join S to the centre of the circle, and draw OS′ perpendicular to CS. We have, then (see fig.), CS : CO′ :: CO : CS′, and, therefore, CS . CS′ = CO . CO′, equal to the square of the radius. It follows that OS′ is the polar of S, and this proves Prop. 1°.

2°. Again, let OS′ represent any right line, passing through a given point, O, whose polar is O′S. Draw CS perpendicular to OS′; we have then, as before, CS . CS′ equal to the square of the radius; and, therefore, S is the pole of OS′, which proves Prop. 2°.

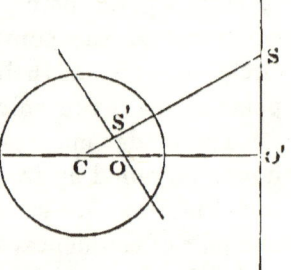

$= A' \cdot ABCD$, both pencils cutting the polar in the same four points, and the Proposition is proved.*

A similar proof applies when the point P is within the circle.

5°. *If four poles be taken on a right line, their polars form a pencil, whose anharmonic ratio is the same as that of the four poles.*

In the first place, it follows from Prop. 1°, that the four polars meet in a point, and therefore form a pencil. Again, it is plain, that the angle subtended by two poles at the centre of the circle, is equal to that made by their polars (as the angle between two lines is equal to that made by two others perpendicular to them); therefore the pencil formed by joining the centre to the four poles, has the same anharmonic ratio (Art. 15) as that formed by the four polars. *Q. E. D.*

POLAR PROPERTIES OF QUADRILATERALS.

41. We shall now give some remarkable polar properties of quadrilaterals inscribed in, or circumscribed to, a circle.

1°. "If a quadrilateral be inscribed in a circle, the intersection of the diagonals and the extremities of the *third diagonal* (see Art. 11) are three points, such that each is the pole of the line joining the other two."

Let BCC'B' be the quadrilateral (see the figure of Prop. 3° of the last Article). In the demonstration of the Proposition referred to, we proved that OO' is the polar of A; in the same way AO' is the polar of O; therefore, the intersection of OO' and AO' is the pole of AO (Art. 40, 1°); this completes the proof of the Proposition.

2°. "If a quadrilateral be circumscribed to a circle, and an inscribed quadrilateral be formed by joining the successive points of contact, the diagonals of the two quadrilaterals intersect in

---

* This Proposition, expressed trigonometrically (see Note on Art. 26), gives the following equation:

$$\frac{\sin \tfrac{1}{2}AD \cdot \sin \tfrac{1}{2}BC}{\sin \tfrac{1}{2}AB \cdot \sin \tfrac{1}{2}CD} = \frac{\sin \tfrac{1}{2}A'D' \cdot \sin \tfrac{1}{2}B'C'}{\sin \tfrac{1}{2}A'B' \cdot \sin \tfrac{1}{2}C'D'}.$$

There are two other equations of a similar form (not independent, however) corresponding to those given in the Note on Art. 16.

the same point, and form an harmonic pencil; and the *third diagonals* of the two quadrilaterals are coincident."

Let PQRS be the circumscribed quadrilateral, and ABCD the inscribed (see fig.;) and let AB and CD meet in O. Then, as AB and CD are (Art. 38) the polars of P and R, the point O is the pole of the

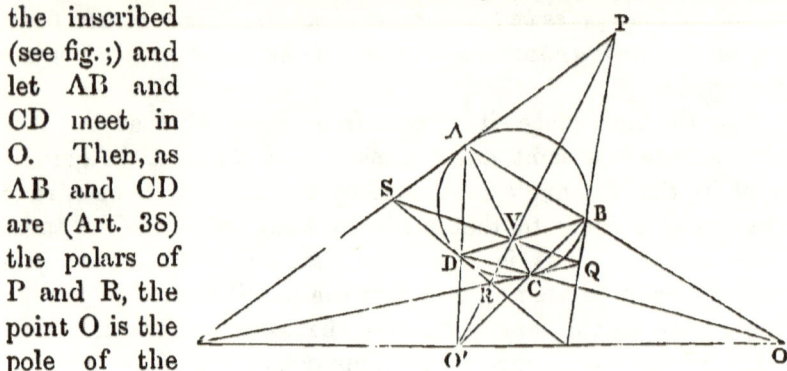

line PR (Art. 40, 1°), and, therefore (Art. 40, 3°), that line passes through V, the intersection of the diagonals of the inscribed quadrilateral; and in a similar way it is proved that the line joining S and Q is the polar of O', and, therefore, passes through V.

Again, as PR has been proved to be the polar of O, it must, when produced, pass through O', the intersection of AD and BC (Art. 40, 3°); but, as O' is the pole of SQ, AO' is cut harmonically (Art. 39); therefore, VA, VS, VD, VR, form an harmonic pencil.

Lastly, as AC and BD are (Art. 38) the polars of the extremities of the third diagonal of PQRS, this third diagonal is the polar of their intersection V (Art, 40, 1°); and therefore is coincident with the third diagonal of the inscribed quadrilateral ABCD (Prop. 1° of this Art).

3°. "If a quadrilateral be circumscribed to a circle, the intersection of each pair of *the three diagonals* is the pole of the remaining one."

Let PQRS be the quadrilateral (see last figure).

In the first place, it is evident from the proof of the last Proposition, that the intersection of PR and QS is the pole of the third diagonal.

It was also proved that O', the intersection of AD and BC, lies on PR, and on the third diagonal of PQRS; therefore, O' is the intersection of PR with the third diagonal; but O' is also the pole of QS; therefore, the Proposition is proved so far as regards the diagonal QS, and in a similar way it can be proved for the remaining diagonal PR.

4°. If a quadrilateral be inscribed in or circumscribed to a circle, whose centre is C, the line joining the centre with the intersection of the diagonals (which point we shall call V), is perpendicular to the third diagonal, and cuts it in a point V', such that $CV \cdot CV'$ is equal to the square of the radius. For, this is merely saying that the point V is the pole of the third diagonal, which has been proved in Props. 1° and 2° of this Article.

5°. "If a quadrilateral be inscribed in a circle, and through the intersection of the diagonals a line be drawn, parallel to the third diagonal, the portion intercepted by either pair of opposite sides is bisected."

Referring to Art. 37, Prop. 3°, let us suppose D and D' to coincide; then, since a chord drawn through any point parallel to the polar of the point, is bisected (Euclid, B. iii. Prop. 3), we have (Art. 36, 2°) the proposition above given.

6°. *Given in magnitude and position one side*, AB (see last figure) *of a quadrilateral inscribed in a given circle, the line joining the intersection of the diagonals to that of the pair of sides adjacent to the given one, passes through a fixed point.*

Let ABCD be the quadrilateral; then, since the points P, V, O', all lie on the polar of O, the line joining V and O' passes through the fixed point P, the intersection of tangents at A and B.

METHOD OF RECIPROCATION.

42. By means of the theory of polars, every Proposition becomes, as it were, double; that is, it leads immediately to another, called its *reciprocal*. The process by which one Proposition is thus deduced from another, is called *reciprocation*. The propriety of the name, as well as the general nature of the process itself, will be understood from the examples which we shall give; but the complete power and extent of the method can only be appreciated in its applications to the higher departments of geometry.

1°. Suppose it were required to find the reciprocal of Pascal's theorem. (See Art. 27.)

Draw tangents at the six vertices of the inscribed hexagon; by this means a circumscribed hexagon is formed, whose three diagonals (that is, lines joining opposite angles) are the polars of the points of intersection of the opposite sides of the inscribed hexagon; for instance, C'F' (see fig.) is the polar of O (Art. 40, 1°). Now, as the three points of intersection are in one right line, their polars pass through one point (Art. 40, 1°). Hence, we have Brianchon's theorem, "*The lines joining the opposite angles of a hexagon, circumscribed to a circle, meet in one point.*"

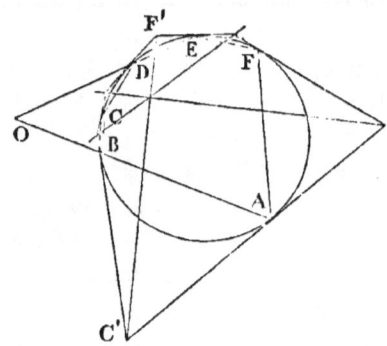

If we examine this process, it will be seen that we have constructed a new figure (the circumscribed hexagon, and its three diagonals), each line in which has for its pole, in respect to the given circle, a corresponding point in the original figure, the tangents having for poles the points of contact. It will also be seen that the original figure stands in exactly the same relation to the new one. The two figures are, therefore, properly called *polar reciprocals;* and the same name is applied to express the relation between their properties. Pascal's and Brianchon's theorems are, then, mutually polar; and either being proved, the other follows as a matter of course.

What we have said of the preceding example may be taken as a general account of the method, which we shall now further illustrate by other examples.

The fundamental properties of polars, proved in Art. 40, 1°, 2°, and 5°, should be constantly borne in mind.

2°. If we suppose the circumscribed hexagon to become successively a pentagon, a quadrilateral, and a triangle, by the coincidence of two points of contact, we have the following theorems, the *reciprocals* of those given in Art. 28:—

" If a pentagon be circumscribed to a circle, and the extremities of any side be joined to those of the two adjacent sides, the intersection of the joining lines is *in directum* with the point of contact on the first side, and the opposite angle of the pentagon."

"If a quadrilateral be circumscribed to a circle, the line joining either pair of opposite points of contact passes through the intersection of the diagonals." (This we have already proved in Prop. 2° of Art. 41.)

"If a triangle be circumscribed to a circle, the lines joining the angles to the opposite points of contact meet in a point."

3°. the anharmonic ratio of four points on a circle is constant (Art. (25); required the reciprocal property.

We begin by drawing the polars of the four given points, A, B, C, D, and of the variable point V (see fig.); that is, we draw five tangents at those points, and thus get four points, A', B', C', D', the poles of the four lines AV, BV, CV, DV. Now, recollecting (Art. 40, 5°) that the anharmonic ratio of four points in a right line, is the same as that of their four polars, it follows that *the anharmonic ratio of the four points, where a variable tangent cuts four fixed tangents to a circle, is constant.* This is the required reciprocal theorem.

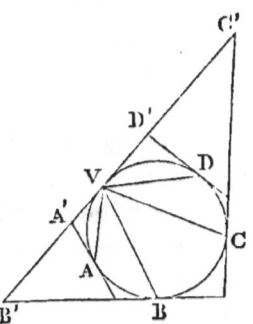

4°. *If tangents be drawn to any plane curve, and their poles be taken with respect to a given circle, the locus of the poles is a new curve, called the reciprocal of the original.* In what does the reciprocity consist?

Let A'T and B'T (see fig.) be two tangents to the original curve A'B', and let A and B be the two corresponding points in the new curve AB; then, T, the intersection of the tangents, is the pole of the chord AB, with respect to the given circle, (whose centre is the point C on the figure). Let us now suppose the point B' to approach indefinitely to A'; the point B will, in consequence, approach to A, the point T will ultimately coincide with A', and the line AB become a tangent at A; that is, *any*

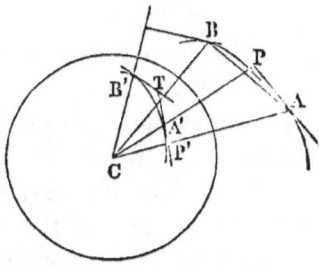

point such as A′ in the original curve, is the pole of the tangent drawn at the corresponding point of the new curve. The reciprocal relation between the two curves is therefore obvious.

It is evident that the lines CA′ and CA are respectively perpendicular to the tangents AP and A′P′, and that CA′ · CP = CA · CP′ = the square of the radius of the circle.

43. In the examples above given, a circle formed part of the original construction. When that is not the case, we may take any circle, and perform the reciprocation with respect to it. The following examples are of this kind:—

1°. A right line meeting the sides and diagonals of a quadrilateral is cut in involution (Art. 37, 2°); what is the reciprocal property?

Take any circle, and construct a new figure, in which each line shall be the polar with respect to the circle of a corresponding point in the original figure. We shall thus have a new quadrilateral, the six intersections of whose sides are the poles of the sides and diagonals of the given quadrilateral. *Corresponding to the six points in involution, we shall have a pencil of six lines in involution* (Art. 40, 1° and 5°), whose vertex is the pole of the transversal, and whose legs pass through the intersections of the sides of the new quadrilateral. The reciprocal theorem is, therefore, that which we have already given in Art. 37, Prop. 1°.

2°. If the angles of a variable triangle move on three given right lines, and if two sides pass respectively through two given points *in directum* with the intersection of the two given lines on which the opposite angles move, the third side will pass through one of two fixed points (Art. 19, 2°). What is the reciprocal?

Take any circle, and construct a new figure as before, having a line corresponding to each point, and a point corresponding to each line in the figure of the given proposition. We shall find the following result, already given in Art. 19, 3°:—

If the sides of a variable triangle pass each through a given point, and if two of the angles move on two given lines, whose intersection is *in directum* with the two points through which the opposite sides pass, the third angle will move on one of two fixed lines.

3°. In like manner the Propositions given in Arts. 21 and 22 are polar reciprocals with respect to any circle.

44. The following theorems are the reciprocals of others already proved:—

1°. "If from the same point be drawn two tangents to a circle, any third line, and a fourth to the pole of the third line, the four lines form an harmonic pencil." (Reciprocal of the Proposition in Art. 39.)

2°. *If a quadrilateral be circumscribed to a circle, two tangents from any point, and four lines from the same point to the angles of the quadrilateral, form a pencil in involution.* This is the reciprocal of Art. 37, 3°. It may be proved independently by the aid of the property established in Art. 42, 3°. We shall give the proof as an illustration of the use of that important property.

Let $ABA'B'$ be the quadrilateral, and V the point. Let the tangents VC, VC' (see fig.) be produced to meet $A'B$. Then considering VC, VC', AB, $A'B'$, as four fixed tangents, the anharmonic ratio of their intersections with a fifth variable tangent is constant; therefore, $V \cdot ACC'B' = V \cdot BCC'A' = V \cdot A'C'CB$, AB' and $A'B$ being taken successively as the fifth tangent.

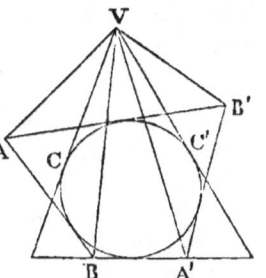

3°. *If three pairs of tangents be drawn to a circle, from three points in a right line, any seventh tangent will be cut in involution.* This is the reciprocal of the Proposition proved in Art. 37, 4°.

4°. We saw in Art. 40, 5°, that a line passing through the intersection of the diagonals of a quadrilateral, inscribed in a circle, and parallel to the third diagonal, gives equal intercepts between that point and either pair of opposite sides. In place of saying that the intercepts are equal, we may say that they subtend equal angles at the centre, which is an evident consequence of the former result. Taking the reciprocal of the theorem in this form, we have the following Proposition:—

If a quadrilateral be circumscribed to a circle, and a perpendicular be drawn from the centre to the third diagonal, lines *from the foot of the perpendicular, to either pair of opposite angles, make equal angles with the third diagonal.* This follows immediately from the former, bearing in mind that the angle between two polars is equal to that subtended at the centre by their poles.

It is not difficult to establish this Proposition independently.

Some additional examples of reciprocation, involving properties of angles, will be found in Chapter IX.

45. In connexion with the method of reciprocation, the following principle, due to Mr. Salmon, is of use:—

The distance of any two poles from the centre of a circle, are to one another as their distances from the alternate polars.

Let O and P be the two poles, and OA and PB (see fig.) their distances from the alternate polars. Draw OX and PY perpendicular to CP and CO respectively. Then, since CO · CO′ = CP · CP′ (both being equal to the square of the radius), CP′ : CO′ :: CO : CP, or, :: CX : CY; therefore, CP′ − CX : CO′ − CY :: CP′ : CO′, or, :: CO : CP; therefore, CO : CP :: OA : PB. Q. E. D.

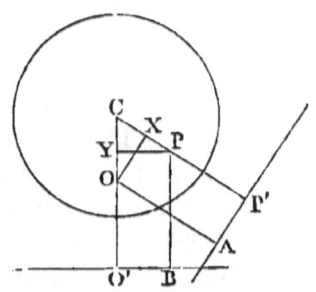

46. In order to exemplify the application of the foregoing principle, we shall first prove the following theorem :—

If a quadrilateral be inscribed in a circle, the rectangles under the perpendiculars, drawn from any point in the circumference to each pair of opposite sides, are equal.

Let ABCD be the quadrilateral, and OP, OP′, OQ, OQ′, the perpendiculars. Join OA, OC, PQ, P′Q′. Then, as the quadrilateral, OQAP, has the angles at P and Q right angles, it is inscribable in a circle, and therefore the angle POQ is the supplement of PAQ; for a similar reason the angle P′OQ′ is the supplement of P′CQ′; therefore, the angle

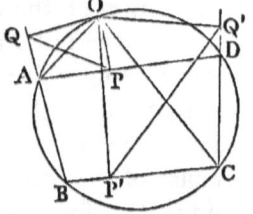

POQ = P′OQ′ (Euclid, B. iii. Prop. 22). Again, the angle OQP = OAP (Euclid, B. iii. Prop. 21) = OCD = OP′Q′; the triangles OPQ and OQ'P' are, therefore, similar, and we have OP : OQ :: OQ' : OP'; therefore, OP · OP' = OQ · OQ'. *Q. E. D.*

If the points A and B become coincident, and also C and D, AB and CD will become tangents. We have then the following theorem :—

If perpendiculars be drawn from any point in the circumference of a circle upon two tangents and their chord of contact, the square of the last perpendicular is equal to the rectangle under the former.

47. Let us now see what theorems can be derived from the preceding by the method of reciprocation.

Draw tangents at the five points A, B, C, D, O; draw LX, NX', perpendicular to the fifth tangent (see fig.); and join the centre C' to O, L, and N. Then, as O is the pole of the fifth tangent, and L the pole of AD, we have (Art. 45) LX : OP :: C'L : C'O; also, NX' : OP' :: C'N : C'O; therefore, LX · NX' : OP · OP' :: C'L · C'N : C'O², that is, in a constant ratio, when the points A, B, C, D are fixed, and O moves on the circumference.

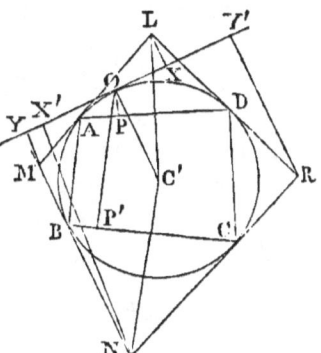

In a similar manner we have MY · RY' : OQ · OQ' in a constant ratio (MY and RY' being perpendiculars from M and R to the fifth tangent, and OQ, OQ', the same as in last Article); and therefore, as OP · OP' = OQ · OQ' (Art. 46), we have finally LX · NX' : MY · RY' a constant ratio. The resulting theorem is as follows :—

If a quadrilateral be circumscribed to a circle, and a fifth variable tangent be drawn, the rectangles under perpendiculars on it from each pair of opposite angles, are in a constant ratio.

If, as before, the points A and B become coincident, and also C and D, we have the following theorem : *If two fixed tangents be drawn to a circle, and also a third variable tangent, and if perpendiculars be drawn on the third tangent from the points of contact of the fixed tangents and their point of intersection, the square of the last perpendicular is to the rectangle under the former in a constant ratio.*

PROBLEMS RELATING TO THE THEORY OF POLARS.

48. We shall now exemplify the use of the theory of polars in the solution of problems.

1°. *From a given point* O, *without a given circle, let a secant be drawn, cutting the circle in* P *and* Q, *and let a portion,* OX, (see fig.) *be taken on it, such that* $\frac{1}{OX} = \frac{1}{OP} + \frac{1}{OQ}$; *required the locus of the point* X?

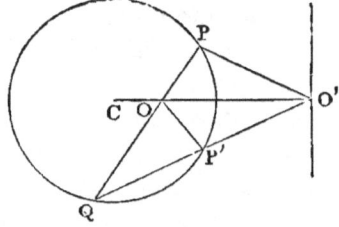

Construct the polar of the given point, and let it cut the revolving secant in the point O'. Then, as the secant is cut harmonically, we have (Art. 5, 3°) $\frac{1}{OO'} = \frac{1}{2}\left(\frac{1}{OP} + \frac{1}{OQ}\right) = \frac{1}{2} \cdot \frac{1}{OX}$; therefore, OO' = 2·OX; the required locus is, therefore, a right line, parallel to the polar, and bisecting OO' in any one of its positions.

Should the point O be within the circle, the result above obtained will still hold good in the sense explained in Art. 13.

If the revolving line meet any number of given circles, or circles and right lines in P, Q, R, S, *&c., and if* $\frac{1}{OX} = \frac{1}{OP} + \frac{1}{OQ} + \frac{1}{OR} +$ *&c., it will appear, as in Art.* 14, *that the locus of* X *is still a right line.*

2°. "Given a circle and a point O, another point O' may be found, whose distances, O'P, O'Q, from the extremities of any chord through O, shall be to one another as the corresponding segments OP, OQ of the chord." (The point O is first supposed within the circle.)

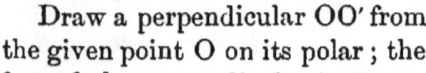

Draw a perpendicular OO' from the given point O on its polar; the foot of the perpendicular is the required point.

For, it appears, from the proof of Art. 39, that when any line is drawn through the pole O, the angle PO'Q is bisected, and, therefore (Euclid, B. vi. Prop. 3), OP : OQ :: O'P : O'Q.

If O be given without the circle, the same result holds good.

Since $CO \cdot CO'$ is equal to the square of the radius, the points O and O' are mutually convertible, so that, if O' were the given point, O would be the point answering the proposed question.

The foot of the perpendicular from a point to its polar may be called the *middle point of the polar*.

3°. "From a given point O, in the produced diameter of a given circle, it is required to draw a secant, cutting the circle in two points, P and Q (see figure), the rectangle $(AP \cdot BQ)$, under whose distances from the adjacent extremities of the diameter shall equal a given quantity."

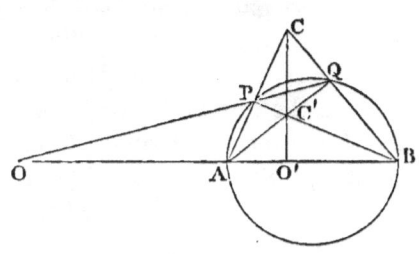

Analysis.—Let OQ be the required secant. Join PB, QA, and produce AP and BQ to meet in C. The point C, and the intersection of AQ and PB, lie on the polar of O (Art. 40, 3°), which is a line given in position, as O and the circle are given. Let CO' be the polar. Now, as $AP \cdot BQ$ is given, the ratio $AP \cdot BQ : AB^2$ is given; that is, the compound ratio $\begin{Bmatrix} AP:AB \\ BQ:AB \end{Bmatrix}$ is given. But, since the triangles APB, AO'C are similar, and also the triangles BQA, BO'C (Euclid, B. iii. Prop. 31), $AP:AB :: AO':AC$ and $BQ:AB :: BO':BC$; therefore (Euclid, B. vi. Prop. 23) the ratio $AO' \cdot BO' : AC \cdot BC$ is given; and therefore (as $AO' \cdot BO'$ is given) the rectangle $AC \cdot BC$ is given. But the difference of the squares of AC and BC is also given, being equal to the difference of the squares of the segments AO', BO' (Euclid, B. i. Prop. 47). The question comes, then, finally to this:—"Given the difference of the squares of two lines, and the rectangle under them, to find the lines." From these conditions, the lines AC and BC may be found (Lardner's Euclid, B. ii. Prop. 14), and the triangle ABC constructed, which evidently determines the position of the required secant.

If the rectangle $PB \cdot AQ$ had been given in place of $PA \cdot QB$, the preceding analysis would again apply, by changing C into

C′, the intersection of PB and AQ, and interchanging the letters P and Q.

The problem may be solved in a similar manner if the point O is given, cutting the diameter internally.

4°. "Given in position one pair of opposite sides of a quadrilateral, inscribed in a circle, and also the point of intersection of the diagonals, to find the locus of the centre of the circle." (See the figure of Art. 40, Prop. 3°.)

Let CA, C′A be the given lines, and O′ the given point, and let O be the intersection of the other pair of opposite sides of any of the inscribed quadrilaterals. Then, as O is the pole of the line joining A and O′ (Art. 40, 3°), A is the vertex of an harmonic pencil, three of whose legs, AC, AC′, AO′, are given in position; and therefore AO, the leg conjugate to AO′, is (Art. 7) also given. A perpendicular from the given point O′ to AO is (Art. 41, 4°) the locus required.

5°. *To inscribe a polygon in a given circle, so that each side shall pass through a given point* (the order of succession of the sides, with respect to the given points, being assigned).

Proceeding as in Article 32, we can construct three polygons, whose sides shall pass through the given points, and all whose angles, except one, shall rest on the circle, this one being the corresponding angle in all. Let the legs containing this angle cut the circle in the points A, D for one of the polygons, in B, E for another, and in C, F for the third, and let K be the point where the corresponding angle of the required polygon should be. It will appear at once, from Art. 40, 4°, that the finding of the point K is reduced to the problem solved in Art. 30; and either of the points so found will determine a polygon answering the question. (See Salmon's Conic Sections, Art. 322.)

As the line which determines the point K (see Art. 30) may not meet the circle, the two solutions above mentioned may become imaginary.

6°. *To circumscribe a polygon to a circle, so that each angle shall rest on a given right line* (the successive order of the angles, with respect to the given lines being assigned).

Analysis.—Join the successive points of contact; we have then an inscribed polygon, each of whose sides passes through a given point (Art. 40, 1°). The problem is, therefore, reduced to the last.

This is an example of the use of reciprocation in the solution of problems.

7°. "To inscribe a polygon in another of the same number of sides, so that each side shall touch a given circle, which touches a pair of consecutive sides of the given polygon."

Construct three polygons, whose sides shall touch the given circles, and all whose angles but one shall rest on the sides of the given polygon, this one being the corresponding angle in all. It will then appear from the property of a fifth tangent, proved in Art. 42, 3°, that the position of the corresponding angle of the required polygon is determined by means of Art. 31.

The solution of this problem is due to Mr. Townsend.

49. Problem 5° of the preceding Article becomes indeterminate in certain cases. Such are the two which follow:—

1°. *When the number of given points is three, each being the pole of the line connecting the other two.*

Let A, O, O' be the three given points (see the figure of Art. 40, Prop. 3°), and let *any triangle*, BCC', be inscribed in the given circle, with two sides, BC, CC', passing through two of the given points, A and O; the third side, BC', will pass through O'.

For, as A is the pole of OO', AC is cut harmonically, and for a like reason (if AO' be produced) OC is cut harmonically; therefore O' is the vertex of an harmonic pencil, of which O'A, O'D, O'C are three legs, and whose fourth leg (conjugate to O'C) passes through the points B and C'; therefore (Art. 7), the points B, O', C', are in one right line.

2°. *When the number of points is even* there is an indeterminate case, which will appear from the following theorem:—

If a quadrilateral, PQRS, *be inscribed in a given circle, so that three of its sides may pass through three given points in a right line, the fourth side will constantly pass through another fixed point in the same right line* (the successive order of the sides being assigned as before).

Let A, B, C be the three given points in a right line; the fourth point, D (see figure on next page), is also fixed.

For, draw QO parallel to the given line; join SO, and let it meet the given line in N. Then, the angle CNS = NOQ = SPQ (Euclid, B. iii. Prop. 22); therefore, the quadrilateral PANS is inscribable in a circle, and CA · CN = CS · CP; but this latter rectangle is a given quantity (Euclid, B. iii. Prop. 36), as the

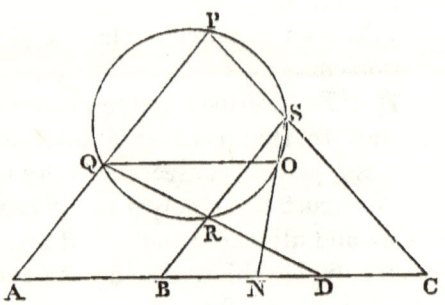

point C and the circle are given; therefore, CA · CN is given, and the point N is determined. Again, the angle RDN = RQO = RSO; therefore, a circle will pass through S, R, N, D (Euclid, B. iii. Prop. 21), and BS · BR = BD · BN; but the first of these rectangles is given; therefore, so is the second, and therefore (as BN is a given line), BD is determined. This proves the theorem.

Let us now suppose a figure of any *even* number of sides to be inscribed in the given circle, *so that all its sides but one shall pass respectively through given points in a right line; the remaining side will also pass through a fixed point in the same right line.*

For, by drawing diagonals from one angle of the polygon it can always be divided into quadrilaterals, to which the preceding theorem becomes applicable. This consideration immediately proves the Proposition.

If the circle and the entire system of points mentioned in this Proposition were given, we could, of course, inscribe an indefinite number of polygons, whose sides would pass respectively through those points.

50. Problem 6°, of Article 48, must also admit of indeterminate cases.

The following correspond to those given in the last Article:—

1°. The problem referred to is indeterminate *when the given lines are three in number, and form a triangle, each of whose sides is the polar of the opposite vertex.*

For, if a triangle be circumscribed to the circle, so that two of its angles shall lie on two of the given lines, its remaining angle will also lie on the third given line. This is the reciprocal of Art. 49, 1°.

2°. The problem is evidently also indeterminate, when the given lines are of an even number, and meet in a point, provided that, if all of them but one be supposed given along with the circle, the remaining one shall coincide with that determined by the following theorem:—

If a polygon of an even number of sides be circumscribed to a given circle, so that all its angles but one shall move respectively on given right lines, which meet in a point, the locus of the remaining angle will be another right line passing through the same point.

This is the reciprocal of the second theorem given in Art. 49, 2°.

51. The problem of inscribing a triangle in a circle, so that each side shall pass through a given point, can be solved in another manner, which we shall explain, as it depends on a very remarkable proposition in the theory of polars.

If two triangles be polar reciprocals, with respect to a circle, (that is, such that every side in each is the polar of a corresponding vertex in the other), the three points of intersection of corresponding sides are in one right line; and the lines joining corresponding angles meet in one point. (In the annexed figure the circle is not drawn.)

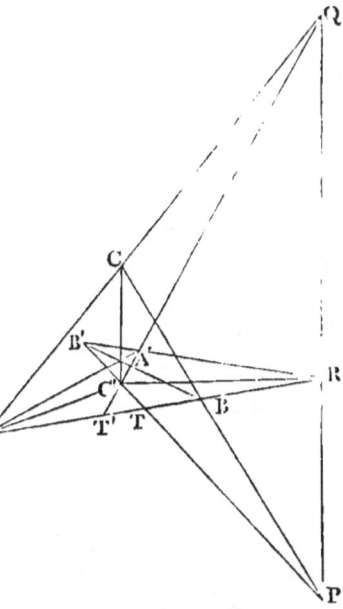

Let ABC, A'B'C', be the two triangles, A being the pole of B'C', A' of BC, &c., and let P, Q, R, be the points of intersection of the corresponding sides. Join C'R, C'A, and AA'. Now, the point of intersection of AB and B'C', which we shall call T (see fig.), is the pole of AC' (Art. 40, 1°), and for a similar reason P is the pole of AA'. We have then four points in a right line, P, T, C′, B′, the poles of the four lines, AA′, AC′, AB, AC, and, therefore (Art. 40, 5°), the anharmonic ratio R . PTC′B′ = A . A′C′T′Q, which again equals R . B′C″TQ = R . QT C′B′. Since R . PTC′B′ = R . QTC′B′, RQ and RP are in one right line (Art. 17); and the first part of the Proposition is proved.

We proved above that P is the pole of AA′, and in the same way it is plain that Q is the pole of BB′, and R that of CC′. Then, as P, Q, R are in one right line, AA′, BB′, CC′ meet in one point. This is the second part of the Proposition, and is the reciprocal of the first part.

52. If we look back to the solution of the problem contained in Art. 48, 5°, we shall see that two triangles (with real or imaginary vertices) may be inscribed in a circle, so as to have their sides passing through the same three points. The six vertices of the two triangles (when real) can (in general) be found by a construction founded on the second part of the preceding Proposition.

Let A′, B′, C′ be the three given points, and ABC the triangle formed by their polars. Draw AA′, BB′, CC′ (which, by the last Article, will meet in a point), and let the points L, M, N, where they meet the sides of the triangle ABC be joined (see fig.). The joining lines will meet the circle in six points, one set of which, P, Q, R (see fig.), will be the vertices of one of the inscribed 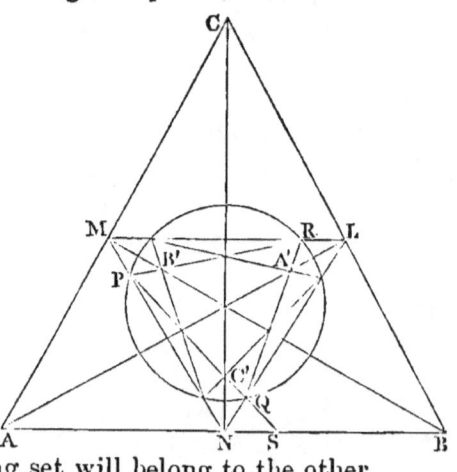 triangles, and the remaining set will belong to the other.

It will evidently be sufficient to prove that the points P, C′, Q are *in directum*, as the same argument will apply to the other vertices.

From Art. 9 it appears that BM is cut harmonically by the lines NC and NL, and, therefore (Art. 7), the line joining P and C′ is (when produced) cut harmonically by NB and NL; but this line is also cut harmonically by the circle (Art. 39), the three points P, C', S remaining unchanged; therefore, it will cut the circle and NL in the same point, that is, it will pass through Q.

This demonstration is taken from Salmon's Conic Sections, Art. 322.

If the three given points were on the circumference, it is easy to see that the two inscribed triangles would become coincident.

If the three points be so situated, that the two triangles A'B'C', ABC, are coincident, the lines AA', BB', CC', become indeterminate, and, therefore, the question itself also, as we have already seen (Art. 49, 1°).*

CHAPTER IV.

THE RADICAL AXIS AND CENTRES OF SIMILITUDE OF TWO CIRCLES.

53. *To find the locus of a point, from which tangents drawn to two given circles are equal.*

Let A and B be the centres of the given circles, and O the point from which the equal tangents OT, OT' are drawn. Draw OP perpendicular to the line joining the centres, and join AT and BT'. Then, since OT and OT' are supposed equal, we have (Euclid, B. i. Prop. 47) $AT^2 - BT^2 = AO^2 - BO^2$. This latter quantity is, therefore, given; and, as AB is

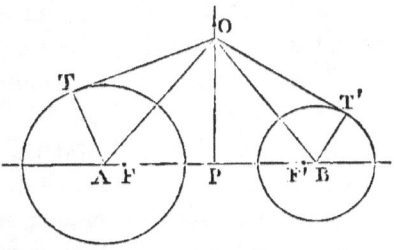

---

* It has been assumed in the foregoing demonstration, that the lines LM, MN, NL will meet the circle, when the problem proposed is possible. This assumption is, however, justified by the proof itself. For, since (as we have seen) those lines (supposing them to meet the circle) determine, by their intersections with it, *all* the vertices that serve to solve the question, if any of them did not meet it, two of the vertices would become imaginary, and consequently the two inscribed triangles also; but this is contrary to the hypothesis.

also given, the locus required is the right line OP perpendicular to AB, and cutting it so that $AP^2 - BP^2 = AT^2 - BT'^2$.*

This right line is called the *radical axis* of the two circles.

When the circles intersect, it is evident (Euclid, B. iii. Prop. 36) *that the radical axis coincides with the common chord.*

In this case, tangents cannot be drawn from a point of the interior portion of the chord. In an algebraical point of view, however, the tangents may still be considered equal, although imaginary.

When the circles touch, the radical axis becomes the tangent drawn at the point of contact. And when they do not meet one another, it meets neither of them. In the last case, which is that represented in the figure, the radical axis is said to be an *ideal chord*, common to the two circles.

If from any point in the radical axis two right lines be drawn, cutting the two circles respectively, the rectangles under the segments of each chord, (measured from the point) are equal (Euclid, B. iii. Props. 35, 36). This property holds good in all the cases above mentioned, and is of importance, as it applies to the points above noticed, from which tangents drawn to the two circles are imaginary.

54. We shall now consider more particularly the case represented in the preceding figure.

Since tangents drawn from P to the two circles are equal, we have $AP^2 - AT^2 = BP^2 - BT'^2$. Now, if we suppose the lengths of the lines BP and BT' to vary simultaneously in such a manner that $BP^2 - BT'^2$ shall remain constant, we shall find an indefinite number of circles, having their centres on the right line AP, any of which may replace the circle whose centre is B, the other circle, whose centre is A, and the radical axis, remaining unaltered. The lengths of BP and BT' are capable of indefinite increase; but it is evident that, as they decrease

---

* We have here assumed the following Proposition (which is evident from Euclid, B. i. Prop. 47):—

When the base and the difference of the squares of the sides of a triangle are given, the locus of the vertex is a right line perpendicular to the base.

simultaneously, BP will be the least possible *when* BT' *vanishes*. This *limiting position of the centre* B is represented in the figure by the point F', which is determined by making PF'^2 equal to the constant quantity $BP^2 - BT'^2$, or $AP^2 - AT^2$.

Let us now conceive the centre and radius of the circle hitherto supposed fixed to vary in a similar manner, that is, without altering the quantity $AP^2 - AT^2$, and it is evident that the radical axis will still remain unchanged, and that there will be a limiting position, F, of the point A, such that PF^2 is equal to the constant quantity $AP^2 - AT^2$. *This limiting point, F, is the position occupied by the centre* A, *when the radius* AT *vanishes, and its distance from the radical axis is equal to that of the other limiting point* F', since $PF^2 = PF'^2$.

It also follows, that *every pair of the entire system of circles found by the variation of the centres* A *and* B *along the line* PA, under the conditions above specified, *will have for their radical axis the line* PO.

The radical axis is obviously common when a system of circles touches the same right line at the same point, or when they pass through the same two points. In the former case, the limiting points coincide with the point of contact. In the latter, they do not exist, or, speaking algebraically, they are imaginary. This appears from the subjoined figure:—

For, the value of PF^2, above given, is $AP^2 - AT^2$, a negative quantity.

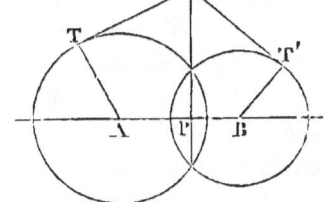

55. If a circle be described with any point, O, on the radical axis, as centre, and one of the equal tangents, OT (see the two preceding figures), as radius, it will touch the two lines AT and BT', and will, therefore, cut the two circles, whose centres are A and B, *orthogonally* (see Art. 6, 1°). And conversely, any circle cutting at right angles two other circles, will have its centre on their radical axis.

From this remark it appears that, *if an infinite system of circles have a common radical axis* (see last Article), *a correlative system also exists*, with the following properties:—

1°. "Every circle of either system cuts orthogonally all the circles of the other system."

2°. "The centres belonging to either system lie on the radical axis of the other."

3°. "Every circle of one system passes through the limiting points (if there be such) of the other system." (This is implied in 1°, since the limiting points are to be regarded as infinitely small circles.)

4°. "If the limiting points of either system be real and distinct, those of the other are imaginary; and *vice versâ*. If in one system they be coincident, the same is true of the other."

56. A system of circles, having a common radical axis, possesses another remarkable property:—

"Take any of the circles which cut the given system orthogonally, and let one of its diameters be drawn, the polar of either extremity of this diameter, with respect to any circle of the system passes through the other extremity."

Let RS be a diameter of the circle cutting orthogonally the system of circles, two of which are represented in the figure, having for centres the points A and B. Let the line joining R and A cut this circle in V, and let the lines SV and AT be drawn. Then, from the conditions of the question, it follows that AT is a tangent to the circle whose centre is O; therefore, $RA \cdot AV = AT^2$, and, as the angle RVS is a right angle, the line SV is the polar of the point R, with respect to the circle whose centre is A; that is, this polar passes through the point S. It is evident that a similar result will hold if the polar of R be taken with respect to any other circle of the system, for instance that whose circle is B. Exactly in the same way it appears that the polars of S will pass through R.

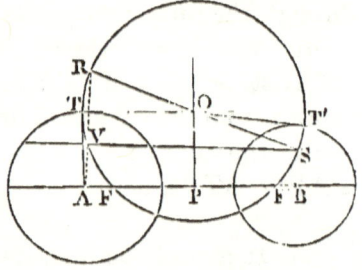

57. This theorem enables us to solve the following problem:—

Given a point, and a system of circles having a common radical axis, it is required to find another point through which the polars of the given point, with respect to the circles of the system, shall all pass.

The question is manifestly reduced to this:—

"Through a given point to describe a circle, so as to cut orthogonally a given system of circles having a common radical axis." This may be done as follows:—

1°. If the given system has real and distinct limiting points, a circle described through them and through the given point, will be the circle required (Art. 55, 3°).

2°. If the limiting points be coincident, we shall have to describe a circle passing through a given point, and touching a given right line at a given point, a problem whose solution is obvious.

3°. If the limiting points be imaginary, that is, if the given system of circles have a common chord, describe a circle through the given point, and through the extremities of the chord; draw a tangent to this circle at the given point, and let it meet the chord produced; a circle described with the point of intersection as centre, and the tangent as radius, will be the required circle (Euclid, B. iii. Prop. 36).

58. When the point given in the preceding problem coincides with either of the limiting points F, F' (see figure in Art. 53), its polar, with respect to any circle of the system, is *constant*.

For, since (Art. 54) $AP^2 - AT^2 = PF^2$, $AT^2 = AP^2 - PF^2 = (AP + PF)(AP - PF) = AF' \cdot AF$; and, in like manner, $BT'^2 = BF \cdot BF'$; consequently, the polar of F, with respect to the circle whose centre is A, and its polar with respect to the circle whose centre is B, are the same line, namely, a perpendicular to AB drawn through F'. Since A and B represent the centres (Art. 54) of any pair of circles of the entire system the Proposition is proved.

It follows from this, that *if two circles be given (not meeting one another), two points can be found, such that each has the same polar with respect to one circle as it has with respect to the other.*

59. We saw in Art. 37, 5°, that a right line cutting a system of circles, which have a common chord, gives a system of points in involution. It will appear from Art. 53, that a similar result

holds good when the common chord is *ideal*. The centre of the system of points is on the radical axis, and the foci are the points where two of the circles touch the line. In the present case, those points are always real, and are found by taking on the transversal a portion on each side of the centre equal to the distance of this point from either of the limiting points (Art. 55, 3°). When the transversal coincides with the line containing the centres of the circles, the limiting points become themselves the foci.

This example of a system of points in involution has been noticed by Mr. Davies in the "Mathematician."

60. The limiting points above-mentioned, considered in relation to the radical axis and a *single* circle of the system, possess some properties worth noticing.

1°. Through either of the limiting points (suppose F), let any right line be drawn, cutting the circle in C and C', and the radical axis in O; then, OC, OF, OC', are proportionals.

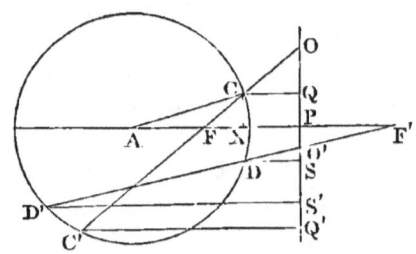

For, since (Art. 55, 3°) OF is equal to the tangent drawn from O, we have $OC \cdot OC' = OF^2$. In like manner O'F' (see fig.) is a mean proportional between O'D and O'D'.

2°. Making the same suppositions as before, let CQ, C'Q' be drawn perpendicular to the radical axis; then, the rectangle $CQ \cdot C'Q'$ is constant.

For, since by Proposition 1°, OC : OF :: OF : OC', it follows (by similar triangles) that CQ : FP :: FP : C'Q'; therefore, $CQ \cdot C'Q' = FP^2 =$ constant.

In like manner the rectangle $DS \cdot D'S'$ (see fig.) $= F'P^2 =$ constant. It is evident that the constant quantity is the same whichever of the points F, F' is taken.

3°. Since PF is equal to the tangent to the circle from P, $AP^2 - PF^2 =$ the square of the radius $= AC^2$ (see fig.). We have then $AF^2 + 2AF \cdot FP = AF^2 + FC^2 + 2AF \cdot FX$ (X being the foot of the perpendicular from C upon AP), and therefore, $2AF \cdot XP = FC^2$, or, $2AF \cdot CQ = FC^2$.

In a similar manner we shall have $2AF' \cdot DS = F'D^2$.

61. The preceding properties may easily be exhibited in a different form. The last, for instance, gives a locus of some importance.

Given a point (F or F'), *and a right line* OP *in position* (see last figure), *required the locus of a point* C, *the square of whose distance from the given point shall equal the rectangle under a given line and the perpendicular from* C *upon the line given in position.*

It follows from the property referred to that the locus is a circle lying entirely on one side of the given right line. If it is supposed to lie on the same side as the given point (F), the circle is constructed by drawing the perpendicular FP upon the given line, and producing it through F, until AF is equal to half the right line given in length. The point A so found is the centre, and the square of the radius is equal to $AP^2 - PF^2$, a given quantity.

If the locus is supposed to be at a side of the given line different to the given point (F'), draw F'P perpendicular to the given line, and produce it through P until AF' is equal to half the right line given in length. The point A so found is the centre, and the square of the radius equals $AP^2 - PF'^2$, a given quantity.

In the latter case the locus vanishes into a point when the line of given length is equal to four times PF', and becomes imaginary when it is less.

THE RADICAL CENTRE.

62. It was remarked by Monge that *the common chords of every pair of three intersecting circles meet in a point.* This is evident *ex absurdo*, from the annexed figure. For, let V be the point of intersection of two of the chords, and, if possible, let the third chord not pass through V. Join CV, and produce it to C' and C'', and we readily find (Euclid, B. iii. Prop. 35) $CV \cdot VC' = CV \cdot VC''$, which is absurd.

The Proposition just proved suggests the more general statement:—

The radical axes of every pair of a system of three circles meet in a point. (This point is called *the radical centre* of the three circles.)

For, if tangents be drawn to the three circles from the point of intersection of two of the radical axes, they are evidently equal, and are, therefore, all imaginary or all real, that is the point of intersection is either within each of the three circles, or without them all (a result easily verified by drawing the figures of the various cases). Now the Proposition has been proved already on the former supposition; and on the latter it follows at once from the equality of the tangents.*

The consideration of the radical centre evidently enables us (Art. 55) *to describe* (when possible) *a circle cutting three given circles orthogonally.*

63. The following theorem, given by Mr. Davies (see the Lady's Diary for 1850), is a good illustration of the foregoing principle:—

If any two circles X *and* Y *be described cutting three given circles,* P, Q, R, *and if two triangles,* ABC, A'B'C', *be formed, whose sides coincide with the common chords found by taking* X *and* Y, *respectively, with the three given circles, the points of intersection of the corresponding sides will lie in one right line.*

For (Art. 62), a pair of corresponding angles (suppose A and A') will lie on the radical axis of the circles Q and R; and therefore the line joining them is the radical axis of Q and R. Hence, the lines joining *every* pair of corresponding angles of the two triangles ABC, A'B'C', meet in a point, namely, the radical centre of the three circles, P, Q, R (Art. 62), and, consequently (Art. 23), the intersections of corresponding sides lie in one right line.

If any of the chords forming the sides of the triangles were *ideal* (see Art. 53), a similar demonstration would apply, and a similar result be arrived at.

64. The solution of the following problem depends on the same principle (see the Lady's Diary for 1851):—

To describe a circle such that the radical axes determined

---

* The Proposition just proved leads immediately to the following:—

If a variable circle be described through two given points to cut a given circle, the common chord passes through a fixed point on the line which joins the given points.

This will appear by taking two positions of the variable circle.

by it and three given circles shall pass respectively through three given points.

Let P, Q, R be the given circles, and X the circle required (see last Article). It follows from what has been said that the three radical axes in question will form the sides of a triangle (such as ABC) whose angles lie on the radical axes belonging to every pair of the given circles. The question proposed comes then to this:—" To describe a triangle, having its angles on three given right lines, and its sides passing each through a given point." In the present case, the three given right lines meet in a point, and the triangle is readily constructed by the aid of Art. 22. The centre of the required circle is found by drawing perpendiculars from the centres of two of the given circles upon the corresponding sides of the triangle, and its radius is determined by the final equation given in Art. 53.

CENTRES OF SIMILITUDE.

65. We shall now explain the properties of the *centres of similitude* of two circles.

" If the line joining the centres A, B of two circles be divided externally at C, and internally at C′, into segments propor-

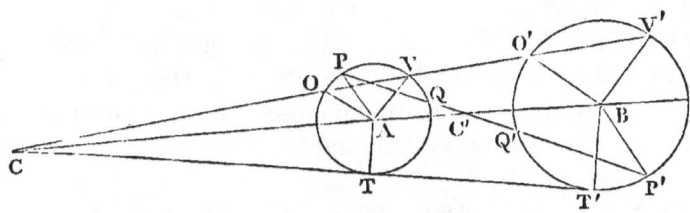

tional to the corresponding radii, the former point of section is said to be the *external*, and the latter the *internal* centre of similitude of the two circles." The propriety of the name will appear from the second of the properties which follow.

1°. A tangent drawn from either centre of similitude (suppose C) to one of the circles, is also a tangent to the other.

Let CT touch the circle whose centre is A. Join AT, and draw a perpendicular from B upon CT (produced, if necessary). Then, since AC : BC :: AT : the perpendicular, it follows from the definition above given, that the perpendicular is equal to the radius of the circle whose centre is B, and therefore the line CT touches that circle.

A common tangent to two circles, such as TT', may be called direct, and one through the internal centre of similitude C' transverse.

It follows from what has been said, that when one of the circles is entirely within the other, both centres of similitude lie within both circles, since the common tangents are in that case all imaginary.

When the circles touch one another, one of the centres of similitude coincides with the point of contact, and when they intersect, the external centre of similitude lies without both circles, and the internal centre within both.

The student is recommended to draw separate figures, corresponding to the various cases just enumerated, and to follow out the properties of the centres of similitude in each.

2°. "If a right line be drawn from a centre of similitude (suppose C), cutting one circle in the points O, V (see last figure), and the other in O', V', we shall have in all cases, CO : CO' :: the radius AO : the radius BO', and CV : CV' in the same ratio." (The points O, O' are supposed to be taken, so that the angles COA and CO'B are both obtuse or both acute.)

For, since CA : CB :: AO : BO', we have CA : AO :: CB : BO', and therefore (Euclid, B. vi. Prop. 7) the triangles CAO and CBO' are similar. In like manner the triangles CAV and CBV' are proved to be similar, and the Proposition immediately follows:—

"If the secant be drawn from the other centre of similitude C', we shall have C'P : C'P' :: the radius AP : the radius BP', and C'Q : C'Q' in the same ratio" (see last figure).

Two points, such as O, O', or V, V', are said to *correspond*. In like manner P, P' and Q, Q' are corresponding points. *Tangents at two such points are evidently parallel*, since the corresponding radii are parallel.

3°. "We shall also have, in all cases, CO · CV' = constant, for all secants drawn from C, cutting the same pair of circles, and CV · CO' = the same constant quantity."

For, $CO \cdot CV' : CO' \cdot CV' :: CO : CO'$, that is, in a constant ratio (Prop. 2°), but $CO' \cdot CV'$ is constant (Euclid, B. iii. Props. 35, 36), therefore, $CO \cdot CV'$ is constant.

Again (Prop. 2°), $CO : CO' :: CV : CV'$, therefore $CV \cdot CO' = CO \cdot CV' = $ constant.

"In like manner, $C'P \cdot C'Q' = C'Q \cdot C'P' = $ constant."

Two points, such as O and V', or V and O', may be said to *correspond inversely*. The same may be said of P and Q', and of P' and Q.

4°. *If a circle touch two others, the right line joining the points of contact passes through their external centre of similitude, when the contacts are of the same kind, and through the internal centre when of different kinds.*

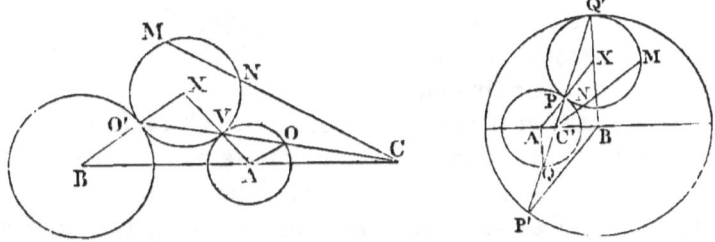

(In the first of the annexed figures, the contacts are of the same kind; and, in the second, of different kinds.)

In the first case, join the points of contact O', V, and let the joining line cut the line joining the centres in C, and draw the radii BO', AV, AO. The radii BO' and AV will meet in X, the centre of the circle touching the two others. The angle VO'X $= $ O'VX $= $ AVO $= $ AOV; therefore, AO and BO' are parallel, and $AC : BC :: AO : BO'$, that is to say, C is the external centre of similitude of the two circles whose centres are A and B.

In the second case, making a similar construction, we find the angle PQ'X $= $ Q'PX $= $ APQ $= $ AQP; therefore, BQ' and AQ are parallel, and (calling C' the point where the line PQ' meets AB) $AC' : C'B :: AQ : BQ'$, that is, C' is the internal centre of similitude of the two circles touched by the circle whose centre is X.

In a similar way the Proposition may be proved for any other variety of the figure.

It is to be observed that *in all cases the points of contact are such as we have called inversely corresponding* (see Prop. 3°).

66. The following applications will serve to illustrate the importance of the properties just proved:—

1°. *It is required to draw a right line parallel to a given one, so that one of the parts intercepted by the circumferences of two given circles may be a maximum.*

Let A and B (see fig.) be the centres of the given circles, and K the given right line. Through C′, the internal centre of similitude, draw a right line, PP′, parallel to K; the intercept PP′ is a maximum.

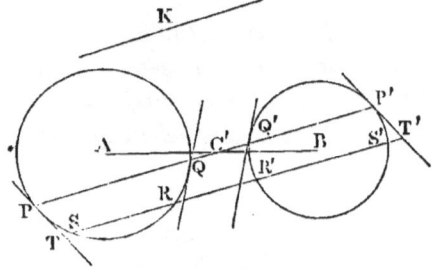

For, let SS′ be the corresponding intercept made by any other right line parallel to K, and let it meet tangents at P, P′ in T, T′. Now, since the tangents (Art. 65, Prop. 2°) are parallel, PP′ = TT′, and, therefore, PP′ is greater than SS′. *Q. E. D.*

It is to be observed that *when each of the given circles is entirely without the other, the same right line PP′ determines a minimum as well as a maximum.*

For, the tangents at Q, Q′ (see fig.) are parallel, and therefore (the circumferences at those points being turned in *opposite* directions) the intercept QQ′ is *less* than the corresponding portion of any other line parallel to K, such as RR′.

2°. *Any number of circles being described to touch two given circles (the contacts being supposed of the same kind), another circle may be found which will cut them all orthogonally.*

Referring to the first figure of Art. 65, 4°, let us suppose the circle whose centre is X to represent one of the indefinite number of circles. Now, since O′, V are points inversely corresponding, the rectangle CV · CO′ is the same for all the circles (Art. 65, 3°); therefore, tangents to them drawn from C are all of equal length, and consequently, a circle described with C as centre, and one of the tangents as radius, will cut them all orthogonally.

It is easy to prove that *the radical axis of the variable circles taken in pairs all pass through* C. If the contacts were of *different* kinds, they would pass through the *internal* centre of similitude of the two given circles, and the circle orthogonal to the system of touching circles would become imaginary.

3°. *It is required to describe a circle through a given point to touch two given circles.*

Referring again to the first figure of Art. 65, Prop. 4°, let us suppose M to be the given point, and MNV to be the required circle. From what was said in that Proposition, it is evident that the rectangle $MC \cdot CN$ (which is equal to $CV \cdot CO'$) is a known quantity; therefore, since M and C are given points, CN is a known length, and consequently the point N is determined. The proposed question is now reduced to the following:—

To describe a circle through two given points to touch a given circle (either of the two original circles may be taken). This may be done as follows:—

Through the given points M, N (see fig.), describe any circle cutting the given circle DTT', and let the common chord, DE, be produced to meet the line MN in F; from F draw two tangents FT, FT'; a circle described through M, N, and either of the points of contact, T or T', will touch the given circle.

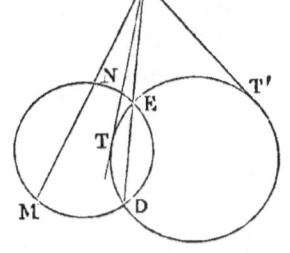

For, the rectangle $MF \cdot FN = DF \cdot FE = FT^2$; therefore, a circle described through M, N, and T, will touch FT at the point T (Euclid, B. iii. Prop. 37), and will, of course, touch the given circle at the same point. The same proof applies to the point T'.

If MN and DE be parallel, the points T and T' are found by drawing a perpendicular to MN at its middle point.*

Returning now to the orignal question, it is evident that it admits of two solutions by the aid of the external centre of

---

* The following important problem can be reduced to that now solved:
"To find a point, C, in a given right line, so that the sum or difference of its

similitude. Two more may be obtained by means of the internal centre, so that the entire number of solutions is four.

AXES OF SIMILITUDE.

67. By the aid of the problem whose solution has just been completed, we can *describe a circle to touch three given circles.* The reduction of this to the former we shall leave as an exercise to the student. There is, however, another method of solution, by which the points of contact of the required circle with the three given ones are directly determined. This method we propose to explain, as it is connected with some remarkable properties of a system of three circles, and also as it will furnish additional applications of the principles established in Art. 65. We shall commence with the following Proposition:—

The six centres of similitude of three circles taken in pairs lie three by three on four right lines (called axes of similitude).

In order to establish this Proposition we shall first prove that the three external centres of similitude lie on one right line. (This we shall call the *external* axis of similitude.)

Let A, B, C be the centres of the three circles, and C', B', A' the external centres of similitude (see fig.). Let R, R', R" represent the radii of the circles. We have then (Art. 65)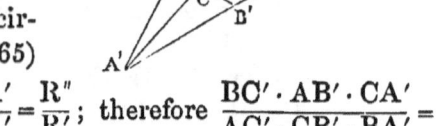

$\dfrac{BC'}{AC'} = \dfrac{R'}{R}$, $\dfrac{AB'}{CB'} = \dfrac{R}{R''}$, $\dfrac{CA'}{BA'} = \dfrac{R''}{R'}$; therefore $\dfrac{BC' \cdot AB' \cdot CA'}{AC' \cdot CB' \cdot BA'} =$

---

distances from two given points, A, B, shall equal a given line."

With A as centre, and the given sum or difference as radius, describe a circle; from the other point, B, draw a perpendicular BP upon the line given in position, and produce it until PE = BP; describe a circle through B and E to touch the former circle; the line joining A to the point of contact, D or D', will determine the required point.

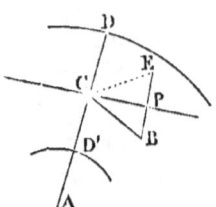

As an example of the use of this problem, we shall give another which is immediately reducible to it:—

"To describe a circle so as to touch two given circles, and have its centre on a given right line."

$\frac{R' \cdot R \cdot R''}{R \cdot R'' \cdot R'} = 1$, and consequently $BC' \cdot AB' \cdot CA' = AC' \cdot CB' \cdot BA'$. It follows from this result (Art. 9, Lemma 2) that A', B', C' are in one right line.

We shall next prove that any of the external centres (such as A') is *in directum* with two internal centres (such as B'', C'',). (There will, in this way, be three right lines, which may be called *internal* axes of similitude.)

In order to see this, it is only requisite to put the letters B'' and C'' (see fig.) in place of B' and C' in the former proof.

68. We shall now give the analysis of the problem referred to at the beginning of the last Article.

A circle touching three given circles may be considered as the limiting case of a circle touching two and passing through a variable point on the third. From this observation, combined with the concluding remark of Art. 66, it appears that the entire number of circles, which can be described to touch three given circles is *eight*. These may be divided into *four pairs*, the circles of each of which have contacts of *opposite* kinds with the given circles. (In the fig. $PP'P''$ and $QQ'Q''$ represent 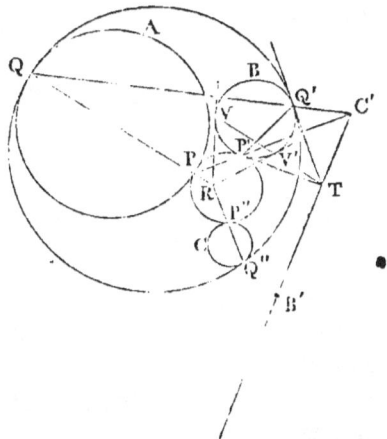 a pair of the required circles, for one of which the contacts are all external, and for the other all internal.) Now (Art. 65 4°), the lines QQ' and PP' meet at C', the external centre of similitude of the two given circles A, B (see fig.); and, also, PQ, $P'Q'$, and $P''Q''$ meet at R, the *internal* centre of similitude of the required pair of circles. Again (Art. 65, 3°), the rectangles $RP \cdot RQ$, $RP' \cdot RQ'$, and $RP'' \cdot RQ''$ are equal, and therefore the point R is the radical centre of the three given circles, and consequently a given point. For the same reason,

K

the rectangles $C'Q \cdot C'Q'$, $C'P \cdot C'P'$ are equal, and therefore C' is a point on the radical axis of the required pair of circles. Hence it immediately follows that this radical axis is the external axis of similitude of the three given circles, and is therefore given. Moreover, if tangents be drawn at the points P', Q', their intersection T must lie on the same radical axis (as the tangents $P'T$ and $Q'T$ are equal), and therefore (Art. 40, 1°) the chord of contact $P'Q'$ must pass through the pole of the axis of similitude with respect to the circle B, that is, through a given point. But this chord also passes through R. It is therefore entirely determined; and, in a similar way, the other chords PQ and $P''Q''$. The three chords being constructed, the two sets of points of contact are found, and therefore the two circles themselves.

The points of contact of the other three pairs of touching circles may be determined, in like manner, by means of the three internal axes of similitude.

CHAPTER V.

ADDITIONAL EXAMPLES ON THE SUBJECTS CONTAINED IN THE FIRST FOUR CHAPTERS.

69. THE present Chapter may be regarded as supplementary to those which have preceded. The arrangement hitherto observed shall be adhered to. The following are a few additional examples on the principles of *harmonic proportion and harmonic pencils*.

1°. *Given the harmonic mean and the difference of the extremes, to find the extremes.*

Referring to the figure in Art. 4, we are given OO' and AB, to find AO' and BO'.

Now, $CO \cdot CO' = CT^2 =$ a given quantity; if, then, we consider CO and CO' as two unknown lines, we have their difference and the rectangle under them, from which they can be found (Lardner's Euclid, B. ii. Prop. 10), and consequently AO' and BO' also.

2°. *Given the sum and the sum of the squares of three lines in harmonic proportion, to find the lines.*

Let X, Y, Z represent the three lines. Subtracting the sum of their squares from the square of their sum, we have the sum of the double rectangles under every pair; that is, $2X \cdot Z + 2 \cdot Y \cdot (X + Z)$ will be a known quantity. But (Art. 5, 1°) $2X \cdot Z = Y \cdot (X+Z)$; therefore $3Y \cdot (X+Z)$ becomes known, and consequently $Y \cdot (X + Z)$. Now, considering Y and $X + Z$ as two lines, we have their sum and the rectangle under them and therefore (Lardners' Euclid, B. ii. Prop. 10) Y and $X + Z$ become separately known. We have then $X + Z$ and the rectangle $X \cdot Z$, from which X and Z can be found, and the question is solved.

3°. *The perpendicular CP on the hypotenuse of a right-angled triangle is an harmonic mean between the segments of the hypotenuse made by the point of contact of the inscribed circle.*

Let O be the centre of the inscribed circle, and OT, OV, OX, perpendiculars from it on the sides. Complete the rectangles CN, CY, TB. Since $OX = OT$, $OX = AY$, and therefore the triangle $OKX = AKY$; and, in like manner, the triangle $OLX = BLZ$; and consequently the rectangle $ON =$ the triangle $ABN =$ the triangle ABC. Hence, $AX \cdot BX$ (which is the same as $AT \cdot BV$, or the rectangle ON) $=$ the triangle ABC, and therefore $2AX \cdot BX = (AX+BX) \cdot$ the perpendicular CP, and (Art. 5, 1°) the Proposition is proved.

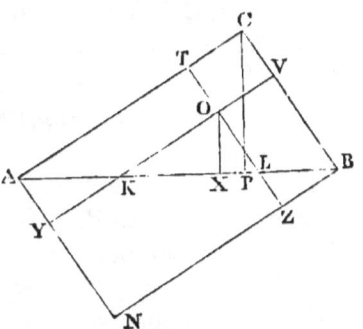

4°. *If two tangents be drawn to a circle, any third tangent is cut harmonically by the two former by their chord of contact, and by the circle.*

Let AO' be the third tangent, cutting the first of the given tangents in A, and the second in B. Draw AV parallel to the second tangent; then, $AO' : BO' :: AV : BT'$ (see fig.). But, since $TX = T'X$, $TA = AV$; therefore $AO' : BO' :: AT : BT' :: AO : BO$. Q. E. D.

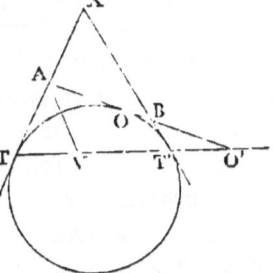

5°. *In a right-angled triangle, the square of the bisector of the vertical angle is an harmonic mean between the squares of the segments into which it divides the hypotenuse.*

Let ABC be the triangle, and CN the bisector of the right angle. Draw OP perpendicular to CN at the point N, and OX parallel to CP. It is evident from the construction that $ON = CN = NP$; therefore $OX = BP$, and $CB : BP :: CB : OX :: CA : AO$. It follows that the rectangles $CA \cdot AO$ and

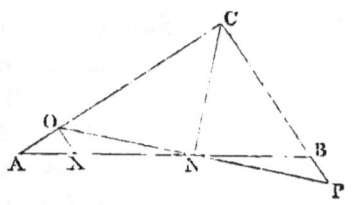

$CB \cdot BP$ (being similar) are in the ratio of the squares of CA and CB (Lardner's Euclid, B. vi. Prop. 20), or as the squares of AN and NB (Euclid, B. vi. Props. 3 and 22). But since the triangles CNO and CNP are isosceles, $CA \cdot AO = AN^2 - CN^2$ (Lardner's Euclid, B. ii. Prop. 6) and $CB \cdot BP = CN^2 - NB^2$; therefore, finally, $AN^2 : NB^2 :: AN^2 - CN^2 : CN^2 - NB^2$. *Q. E. D.*

6°. *If the sides of a triangle be produced, the bisectors of the external angles meet the opposite sides in three points, which lie on one right line.*

Let ABC be the triangle, and let the bisectors meet the opposite sides produced in P, Q, R. Join A and P to V the point where BQ and CR intersect. Then, since V is the centre of the exscribed circle touching BC, the line AV bisects the angle BAC, and therefore (Lardner's Euclid, B. vi. Prop. 3) AC, AV, AB, AP form an harmonic pencil, and consequently VC, VA, VB, VP also form an harmonic pencil (both pencils cutting the line PC in the

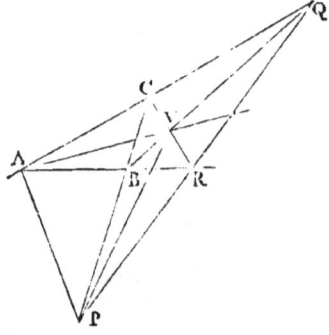

same points). It follows (Art. 7) that the line joining P, R must, when produced, cut AQ and VQ in the same point, namely, the harmonic conjugate of the point R; that is, this joining line must pass through Q, which was to be proved.

This Proposition may also be proved by the converse of Lemma 2° of Art. 9.

7°. It is evident from Art. 9, Lemma 1, that *the lines drawn from the angles of a triangle to the middle points of the opposite sides meet in a point.*

It is easily proved also that they *trisect* each other. For, if AE and BD be the bisectors of two of the sides of the triangle ABC, DE is parallel to AB, and therefore BO : OD :: AB : DE :: AC : CD :: 2 : 1. This Proposition is required in our next example, which is as follows:—

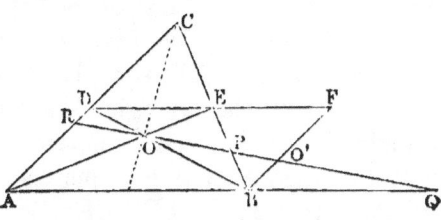

If through the intersection O of the bisectors of the sides of a triangle ABC, a transversal RQ be drawn, cutting the three sides in R, P, Q, it is required to prove that $\frac{1}{OR} = \frac{1}{OP} + \frac{1}{OQ}$, OR *being the intercept which (measured from O) lies in a direction different to that of the other two.*

For, draw BF parallel to AC, and let it meet DE produced in F and the transversal in O'. Now, since CE=EB, DE=EF, and therefore (Art. 7) BD, BE, BO', BQ form an harmonic pencil, and consequently OQ is cut harmonically. We have, therefore, (Art. 5, 3°) $\frac{1}{OP} + \frac{1}{OQ} = 2 \cdot \frac{1}{OO'} = \frac{1}{OR}$ (since OO' : OR :: BO : OD :: 2 : 1).

ADDITIONAL EXAMPLES ON TRANSVERSALS.

70. We shall now give some additional Propositions on the subject of *transversals*.

1°. *Let a right line be drawn parallel to the base of a triangle, and its intersections with the sides be joined to the opposite extremities of the base, to find the locus of the intersection of the joining lines.*

Let the line joining C' and B' (see fig. of Art. 9, Lemma 1) be supposed parallel to the base BC. We have then (Euclid, B. vi. Prop. 2) BC' : AC' :: B'C : AB', therefore BC' · AB' = AC' · B'C, and consequently (Art. 9, Lemma 1) BA' = A'C. The locus required is therefore *the right line joining the vertex A to the middle of the base.*

2°. The principle established in the preceding example is important. As an instance of its application we shall take the following problem:—

"Given all the angles of a quadrilateral and one of its sides, to construct it so that one of its diagonals may bisect the other."

Let AC'OB' (see fig. of Art. 9, Lemma 1) be the required quadrilateral, and AB' the given side; then, since the angles are also given, the triangle AB'B is entirely known, and the triangle AC'C is given *in species* (that is, we can make one similar to it). Now if we construct a triangle of this species, and equal in area to AB'B (Euclid, B. vi. Prop. 25), and if we conceive AC'C to be made equal in all respects to the triangle so constructed, and to be placed with regard to the given triangle AB'B as in the figure, the quadrilateral so found will solve the question. For, if B'C' and BC be drawn, we have the triangle BC'B' = the triangle CB'C'; therefore B'C' is parallel to BC, and consequently (Prop. 1°) BC is bisected by AO; therefore, so is B'C' also.

If the side were not given, it is easy to see that we could find the species of the quadrilateral by taking any line to represent AB', and proceeding as above. The species being determined, any other quantity, such as the area, or perimeter, &c., being supposed given, the problem becomes determinate, and its solution evident.

3°. *If three right lines, AA', BB', CC', be drawn from the angles of a triangle to meet in O, and a new triangle A'B'C' be formed by joining their intersections with the opposite sides, it is required to prove that three right lines, AP, BQ, CR, drawn from A, B, C to the middle points of the corresponding sides of the latter triangle, meet in a point.*

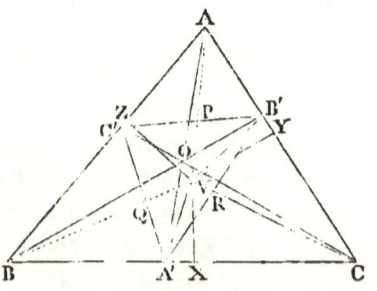

In order to prove this theorem, let us suppose AP and BQ to meet in V; let VX, VY, VZ be drawn perpendicular to the sides of the triangle ABC, and VA', VB', VC' be joined. Then, since QA' = QC', the triangle BA'V = BC'V, that is, BA' · VX = BC' · VZ or $\dfrac{BA'}{BC'} = \dfrac{VZ}{VX}$; in like manner $\dfrac{AC'}{AB'} = \dfrac{VY}{VZ}$; and therefore

$\dfrac{BA' \cdot AC'}{AB' \cdot BC'} = \dfrac{VY}{VX}$. But (Art. 9, Lemma 1) $\dfrac{BA' \cdot AC' \cdot CB'}{BC' \cdot AB' \cdot CA'} = 1$.

It follows that $\dfrac{VY}{VX} \cdot \dfrac{CB'}{CA'} = 1$; that is, $VY \cdot CB' = VX \cdot CA'$, and therefore (joining CV) the triangle $CVB'=$ the triangle CVA'. Hence it appears, *ex absurdo*, that the joining line CV must pass through R, the middle point of A'B'. The Proposition is therefore proved.

4°. The following extension of Art. 9, Lemma 1, is due to Poncelet. (Traité des Propriétés Projectives, p. 85.)

"If lines be drawn from the angles of a polygon of an *odd* number of sides to meet in a point, the continued products of the alternate segments so formed on the opposite sides are equal."

(The figure represents a pentagon, but the proof will apply in general.)

Let O be the point through which the transversals AA', BB', &c. are drawn, and let the lines AC, AD, &c. be drawn joining each angle to the extremities of the opposite side. We have then $\dfrac{AD'}{BD'} = \dfrac{\text{the triangle AOD}}{\text{the triangle BOD}}$, $\dfrac{BE'}{CE'} = \dfrac{BOE}{COE}$, $\dfrac{CA'}{DA'} = \dfrac{AOC}{AOD}$, $\dfrac{DB'}{EB'} = \dfrac{DOB}{EOB}$, and $\dfrac{EC'}{AC'} = \dfrac{EOC}{AOC}$; therefore, by compounding the ratios, $\dfrac{AD' \cdot BE' \cdot CA' \cdot DB' \cdot EC'}{BD' \cdot CE' \cdot DA' \cdot EB' \cdot AC'} = 1$, which proves the Proposition.

Poncelet's demonstration depends on a principle of projection which the reader will find in Chapter VII. of this work.

5°. The second Lemma of Art. 9 is only a particular case of the following general theorem due to Carnot. (Géométrie de Position, p. 295.)

If a transversal be drawn cutting the sides of a polygon (or the sides produced), the continued products of the alternate segments so formed on the sides are equal.

Suppose the polygon to be for instance, a pentagon (see figure on next page), and let C'D' be the transversal. Draw AP, BQ,

&c., perpendiculars from the angles of the polygon on the transversal, and we have
$\dfrac{AD'}{BD'} = \dfrac{AP}{BQ}$, $\dfrac{BE'}{CE'} = \dfrac{BQ}{CR}$,
$\dfrac{CA'}{DA'} = \dfrac{CR}{DS}$, $\dfrac{DB'}{EB'} = \dfrac{DS}{ET}$,
and $\dfrac{EC'}{AC'} = \dfrac{ET}{AP}$; therefore, by compounding the ratios, $\dfrac{AD' \cdot BE' \cdot CA' \cdot DB' \cdot EC'}{BD' \cdot CE' \cdot DA' \cdot EB' \cdot AC'} = 1$. *Q. E. D.* (In the proof C'D' and CD are supposed to meet at a point A'.)

A similar demonstration evidently applies to a polygon of any number of sides.

It is also easy to see that the proof holds good, even when the sides of the Polygon *do not lie in one plane* (in which case the polygon is said to be " gauche"), and the segments are made by a *transverse plane* instead of a right line. We have only to consider AP, BQ, &c. to be drawn perpendicular to the transverse plane, and the proof goes on as before.

6°. *If two transversals,* MN *and* PQ, *cut the opposite sides of a* " gauche" *quadrilateral* ABCD *in such a manner that* $\dfrac{AM}{BM} = K \cdot \dfrac{DN}{CN}$, *and* $\dfrac{AQ}{DQ} = K \cdot \dfrac{BP}{CP}$, K *being a number, it is required to prove that the two transversals meet one another.*

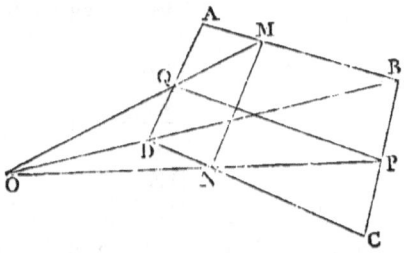

For, the conditions of the question give $K = \dfrac{AM \cdot CN}{BM \cdot DN}$, and $K = \dfrac{AQ \cdot CP}{DQ \cdot BP}$; therefore $AM \cdot CN \cdot DQ \cdot BP = AQ \cdot CP \cdot BM \cdot DN$, or $\dfrac{AM \cdot DQ}{AQ \cdot BM} = \dfrac{CP \cdot DN}{CN \cdot BP}$. Now, since ABD is a triangle, MQ and BD are in the same plane, and may be supposed to meet at a point O, which cuts BD so that $\dfrac{DO}{BO} = \dfrac{AM \cdot DQ}{AQ \cdot BM}$ (Art. 9, Lemma 2);

and for a like reason PN will cut BD in a point, which we may call O′, such that $\frac{DO'}{BO'} = \frac{CP \cdot DN}{CN \cdot BP}$. It follows then that MQ and PN intersect BD in the same point $\left(\text{since } \frac{DO}{BO} = \frac{DO'}{BO'}\right)$, and are therefore in the same plane, and consequently that MN and PQ are also in the same plane. Q. E. D. (Chasles, "Aperçu Historique," p. 242.)

71. The following properties of a complete quadrilateral are worth noticing. (See Art. 11.)

1°. "In *a complete quadrilateral*, the triangle whose bases are any two of the *three diagonals*, and whose common vertex is one extremity of *the remaining diagonal*, are to one another as the triangles having the same bases as the former triangles, and their common vertex at its other extremity." (Carnot, Géométrie de Position, p. 283.)

For (see last figure of Art. 9), we have the triangle ABC : OBC :: AA′ : OA′ and the triangle AC′B′ : OC′B′ :: AX : OX (X being the point where AO cuts B′C′); but since (Art. 9) AA′ is cut harmonically, AA′ : OA′ :: AX : OX; therefore the triangle ABC : OBC :: AC′B′ : OC′B′, and therefore ABC : AB′C′ :: OBC : OB′C′. This proves the proposition when BC and B′C′ are taken as bases, and in a similar manner it may be proved for any other pair of the three diagonals AO, B′C′, BC.

2°. *The middle points of the three diagonals of a complete quadrilateral are in one right line.*

Let IFGM, IEGL, FEML be the component quadrilaterals, and IG, FM, LE the three diagonals of the complete quadrilateral (see figure). Produce these to form the triangle ABC, and join BG and EC. Then, since (Art. 11) BM is cut harmonically, BG also is (Art. 7) cut harmonically; and therefore (since (Art.

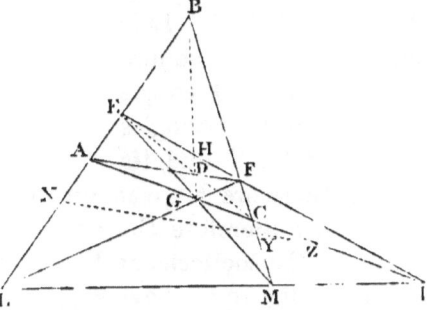

11) BL is cut harmonically) the points A, D, F are in one right line. Now let X, Y, Z be the middle points of LE, FM, IG, and we have (Art. 3) $BX : EX :: EX : AX$; therefore (conversion and alternation) $BE : AE :: BX : EX$, whence, $BE^2 : AE^2 :: BX^2 : EX^2 :: BX : AX$, that is $\dfrac{BE^2}{AE^2} = \dfrac{BX}{AX}$. In like manner we get $\dfrac{AG^2}{GC^2} = \dfrac{AZ}{CZ}$, and $\dfrac{CF^2}{BF^2} = \dfrac{CY}{BY}$, and therefore, by compounding the ratios, $\dfrac{BE^2 \cdot AG^2 \cdot CF^2}{AE^2 \cdot CG^2 \cdot BF^2} = \dfrac{BX \cdot AZ \cdot CY}{AX \cdot CZ \cdot BY}$. The quantity on the left hand equals unity (Art. 9, Lemma 1) (since AF, BG, CE meet in a point D); therefore $BX \cdot AZ \cdot CY = AX \cdot CZ, \cdot BY$ which proves (Art. 9, Lemma 2) that X, Y, Z are in one right line.

This demonstration is borrowed from Poncelet (Traité des Propriétés Projectives, p. 87). There is, however, another demonstration of the property in question which depends on a very useful principle not yet explained. This we shall give in the following Article.

72. LEMMA 4.—*Given in magnitude and position the bases of two triangles having a common vertex; given also the sum of their areas; required the locus of the common vertex.*

Let AB, CD be the bases of the triangles, and V the vertex. Produce the bases to meet in O, and take $OP = AB$, and $OQ = CD$, and join PV, QV, OV, PQ. It is evident then that the quadrilateral OPVQ equals the sum of the triangles ABV and CDV, and its area is therefore a given quantity. But the triangle OPQ is also given; therefore the area PQV is a given quantity, and consequently the locus of V is *a right line* parallel to PQ.

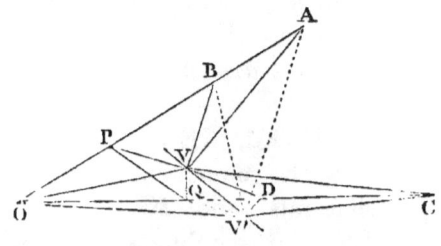

In the figure we have supposed the point V to be between the lines AB, CD. If it should take a position such as V' (see fig.) on the parallel above determined, it is easy to see that it is the *difference* of the triangles ABV' and CDV', which is equal to the given quantity. (The algebraical explanation of this circumstance is, that the triangle CDV changes its sign when its vertex passes from one side of the base to the other.)

We shall now apply this Lemma to prove the second Proposition of the last Article.

Let ABCD be one of the component quadrilaterals (see fig.), and Z, Y, X the middle points of the three diagonals of the complete quadrilateral. Then, since AC is bisected at Z, it is evident that the sum of the triangles ABZ and CDZ $= \frac{1}{2} \cdot$ (quadrilateral ABCD), and since BD is bisected at Y, the sum of the triangles ABY and CDY also $= \frac{1}{2} \cdot$ (quadrilateral ABCD).

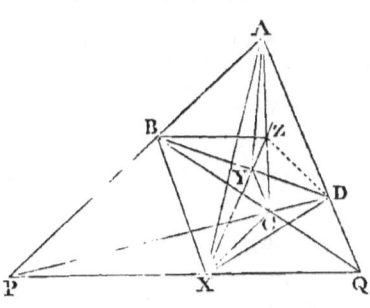

Again, the triangle ABX $=$ APX $-$ BPX $=$ (since PQ is bisected at X) $\frac{1}{2}$ (APQ $-$ BPQ) $= \frac{1}{2} \cdot$ (ABQ), and in like manner the triangle CDX $= \frac{1}{2} \cdot$ (CDQ); therefore the triangle ABX $-$ the triangle CDX $= \frac{1}{2} \cdot$ (ABQ $-$ CDQ) $= \frac{1}{2} \cdot$ (quadrilateral ABCD). It appears now from the Lemma that the three points Z, Y, X, being three positions of the common vertex of two triangles having two fixed bases, AB, CD, and the sum of the areas constant (in the sense above explained), lie on the same right line. Q. E. D.

ADDITIONAL EXAMPLES ON ANHARMONIC RATIO.

73. The following Propositions relate to the application of *anharmonic* principles:—

1°. *If all the sides of a polygon pass through given points which lie in one right line, and all its angles except one move on given right lines, the locus of the free angle is a right line* (the successive order of the sides and angles of the variable polygon, with respect to the given points and right lines, being supposed to be assigned.)

Let us conceive a quadrilateral whose sides pass respectively through four given points, P, Q, R, S, in a right line, and three of whose angles, A, B, C, move on three given right lines. By taking three positions of the quadrilateral, and proceeding as in

Art. 21 (which is a particular case of the present theorem), the student will find that the three positions of the free angle D lie in a right line; and a similar proof will apply in the case of any polygon.

Calling A′, B′, C′ the points where the given right lines, on which the angles (of the quadrilateral) move, are intersected by the right line containing the given points, and calling D′ the point where the locus cuts the same right line, there exists a remarkable relation amongst the segments of D′A′ made by P, of A′B′ made by Q, of B′C′ made by R, and of C′D′ made by S.

In order to see this relation let us consider PS as a transversal cutting the sides of the quadrilateral, and we have (Art. 70, 5°) $PA \cdot QB \cdot RC \cdot SD = PD \cdot QA \cdot RB \cdot SC$. This equation is true for all positions of the moving quadrilateral, and therefore is true in the extreme case, where its sides become *in directum*, and its angles A, B, C, D coincide with A′, B′, C′, D′ respectively. When this takes place, the equation becomes $PA' \cdot QB' \cdot RC' \cdot SD' = PD' \cdot QA' \cdot RB' \cdot SC'$, which expresses the relation in question. (Poncelet, Traité des Propriétés Projectives, Art. 501.) A similar relation obviously holds good in the case of any polygon.

2°. *If all the sides of a polygon pass through given points, and all its angles except one move on given right lines meeting in a point, which is* in directum *with the points through which the sides containing the free angle pass, the locus of this angle is a right line* (the same supposition being made as in the last Proposition.)

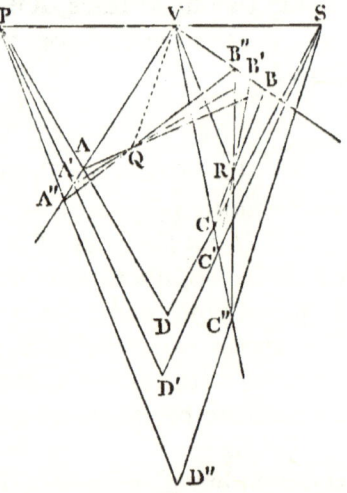

The figure represents the case of a quadrilateral ABCD, in which D is the free angle, and P, Q, R, S the given points. Then, taking three positions of the quadrilateral, and joining Q and R to V, the intersection of the given lines VA, VB, VC, we have the anharmonic ratio $P \cdot VAA'A'' = Q \cdot VAA'A'' = Q \cdot VBB'B'' = R \cdot VBB'B'' = R \cdot VCC'C'' = S \cdot VCC'C''$, and therefore (Art. 20) D, D', D'' are in one right line. The locus of D is therefore a right line.

It is evident that a similar proof will apply when the polygon has any number of sides.

Proposition 3° of Art. 19 is a particular case of this theorem.

74. The *polar reciprocals* (see Art. 42) of the preceding theorems, (with respect to any circle) are the following:—

1°. *If all the angles of a polygon move on given right lines meeting in a point, and all its sides except one pass through given points, this side also constantly passes through a fixed point* (the supposition of the former enunciation being still retained).

The theorem in Art. 22 is a particular case of this.

2°. *If all the angles of a polygon move on given right lines, and all its sides except one pass through given points which lie in a right line passing through the intersection of the lines on which the extremities of the free side move, then this side also passes through a fixed point* (on the same supposition as before).

Proposition 2° of Art. 19 is a particular case of this theorem.

75. The following are additional examples on the application of anharmonic properties:—

1°. *Let a secant* AD *be drawn from the intersection* A *of two tangents* AT, AT' *to a circle and let it cut the circle in two points* B *and* D, *it is required to prove that the rectangle* DT · BT' $= \frac{1}{2} \cdot$ BD · TT'.

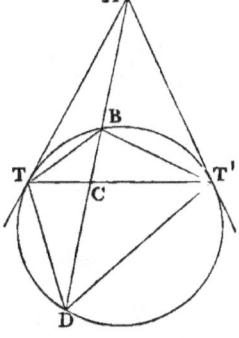

Considering T as the vertex of a pencil whose legs pass through T, T', B and D, we have the anharmonic ratio T · ACBD $=$ T · TT'BD. But since AD is cut harmonically (Art. 39), T · ACBD (which equals $\dfrac{\text{AD} \cdot \text{CB}}{\text{AC} \cdot \text{BD}}$) $= \frac{1}{2}$ (Art. 16); therefore, T · TT'BD also $= \frac{1}{2}$; that is (Art. 26) $\dfrac{\text{TD} \cdot \text{BT'}}{\text{TT'} \cdot \text{BD}} = \frac{1}{2}$, which was to be proved.

This Proposition admits of an easy proof on other principles.

2°. *From two given points* P, R *in the circumference of a given circle to inflect two chords* PO, QO (see fig.), *so that the segments* AC, AB, *intercepted on a given chord* AD, *may have a given ratio.*

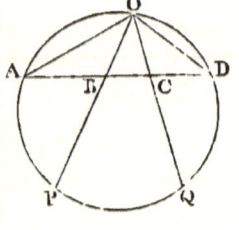

Analysis.—Join AO and DO. Then since A, P, Q, D are given points, $\dfrac{AD \cdot BC}{AB \cdot CD}$ is a given ratio (Art. 25); but $\dfrac{BC}{AB}$ is a given ratio (since AC : AB is given); therefore $\dfrac{AD}{CD}$ is a given ratio, and consequently CD is determined, which solves the question.

3°. *Given a triangle* ABC (see fig.), *let* PQ *and* RS *be drawn parallel to given right lines; required the locus of the intersection of* PB *and* RA, *the segments* AQ *and* BS *being taken in a given ratio.*

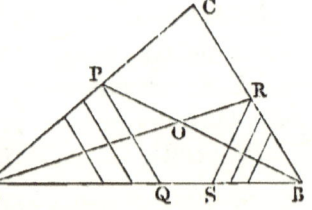

Taking three positions of the point O, whose locus is required, it is easy to see from the conditions of the question, that A and B are the vertices of two pencils having the same anharmonic ratio, and therefore the locus of O is (Art. 20) a right line.

4°. *If, in the last question, the lines* PQ *and* RS *are to pass through given points in place of being parallel to given lines,* the rest remaining as before, it will be found on similar principles that *the locus is still a right line.*

The last question may be considered a particular case of the present, namely, when the given points are at an infinite distance.

ADDITIONAL EXAMPLES ON INVOLUTION.

76. We shall place here two remarkable properties of *six points in involution* not given in the former Articles on that subject.

1°. *The rectangles under the distances from one of the points to the other pairs of conjugates are to one another as the rectangles under the distances from the conjugate of that point to the same pairs of conjugates respectively.*

Let AA', BB', CC', be the three pairs of points. We have to prove that $\dfrac{AB \cdot AB'}{AC \cdot AC'} = \dfrac{A'B \cdot A'B'}{A'C \cdot A'C'}$, and so on.

For, (Art. 35), the anharmonic ratio of the four points A, C, A', B being equal to that of their four conjugates, we have $\dfrac{AB \cdot CA'}{AC \cdot A'B} = \dfrac{A'B' \cdot C'A}{A'C' \cdot AC'}$, and therefore $AB \cdot A'C \cdot A'C' \cdot AB = AC \cdot A'B \cdot A'B' \cdot AC'$, from which the required result follows immediately.

We saw in Art. 37, Prop. 3°, that a right line cutting a circle and the sides of an inscribed quadrilateral is cut in involution. Hence the result above mentioned expresses a property of the circle, A, A' being supposed to represent the points where the transversal cuts one pair of opposite sides C, C', those where it cuts the other pair, and B, B' those in which it intersects the circle. This property is due to Desargues, who proved it for any conic section. The relation between the six points of the transversal he expressed by the phrase "involution of six points," which became one of *five* when a pair of conjugates happened to coincide, as for instance when the transversal became a tangent to the circle. The definition of involution, which we have given in Art. 35, is that laid down by Chasles. (Aperçu Historique, p. 318.)

2°. *If we take three points belonging to three pairs* (suppose A, B, C), *each of them makes two segments with the conjugates of the other two; of the six segments thus formed, the product of three which have no common extremity is equal to the product of the remaining three.*

We have to prove $AB' \cdot CA' \cdot BC' = A'B \cdot C'A \cdot B'C$.

For, (Art. 35), the anharmonic ratio of the four points A, B', C, C' being the same as that of A', B, C', C, we have $\dfrac{AC' \cdot B'C}{AB' \cdot CC'} = \dfrac{A'C \cdot BC'}{A'B \cdot C'C}$, from which the required result is evident.

If in forming the six segments we take the point B' (suppose) in place of B, we have only to interchange the letters B and B' to get the corresponding result, which is $AB \cdot CA' \cdot B'C' = A'B' \cdot C'A \cdot BC$; and so on.

The relation established in this Proposition was known to Pappus as a property of the six points in which a transversal is cut by the sides and diagonals of a quadrilateral. (See Art. 37, 2°.)

It is evident from the proofs given in the present Article, that if three pairs of points in a right line have either of the relations expressed in Propositions 1° and 2°, they are such that the anharmonic ratio of four is the same as that of their four conjugates, and are consequently in involution (Art. 35). Hence either of the relations amongst three pairs of points remarked by Pappus and Desargues implies the other.*

ADDITIONAL EXAMPLES ON THE THEORY OF POLARS.

77. The following properties of a quadrilateral inscribed in a circle are connected with the theory of *polars*:—

1°. "If a quadrilateral be inscribed in a circle, the square of the third diagonal is equal to the sum of the squares of the tangents from its extremities.

Let E, F be the extremities of the third diagonal (see fig.). Now (Art. 41, 1°) TT′, the polar of E, passes through F, and therefore (Lardner's Euclid, B. ii. Prop. 6) $EF^2 - ET^2 = FT \cdot FT' =$ the square of the tangent from F. Hence EF^2 = the sum of the squares of the tangents from E and F.

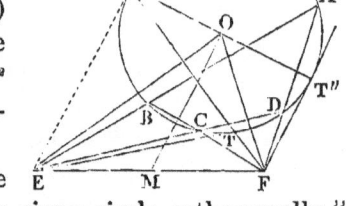

2°. "A circle described on the third diagonal as diameter cuts the given circle orthogonally."

---

* It follows from the Note on Art. 16, that *certain relations amongst the segments of a line cut in involution are true of the sines of the angles made by the legs of a pencil in involution.* (See Art. 37, 1°.) Hence it appears that *if three right lines passing through the same point cut the circumference of a circle in three pairs of points,* A, A′, B, B′, C, C′, (see fig. of Art. 37, 4°), *the relations above mentioned will hold amongst the sines of the halves of the intercepted arcs.* We shall have, for instance, the following equations corresponding to the relations of Desargues and Pappus given in the text:

1°. $\dfrac{\sin\frac{1}{2} AB \cdot \sin\frac{1}{2} AB'}{\sin\frac{1}{2} AC \cdot \sin\frac{1}{2} AC'} = \dfrac{\sin\frac{1}{2} A'B \cdot \sin\frac{1}{2} A'B'}{\sin\frac{1}{2} A'C \cdot \sin\frac{1}{2} A'C'}$

2°. $\sin\frac{1}{2} AB' \cdot \sin\frac{1}{2} CA' \cdot \sin\frac{1}{2} BC' = \sin\frac{1}{2} A'B \cdot \sin\frac{1}{2} C'A \cdot \sin\frac{1}{2} B'C.$

ADDITIONAL EXAMPLES ON THE THEORY OF POLARS. 81

Let M (see last figure) be the middle point of the third diagonal EF, and let it be joined to the centre O. We have then (Lardner's Euclid, B. ii. Prop. 14) $2 \cdot OM^2 + 2 \cdot EM^2 = EO^2 + FO^2 = ET'^2 + FT'^2 + 2 \cdot OT'^2 = $ (Prop. 1°) $EF^2 + 2 \cdot OT'^2 = 4 \cdot EM^2 + 2 \cdot OT'^2$. Hence, $OM^2 = EM^2 + OT'^2$; therefore $EM^2 = OM^2 - OT'^2 = $ the square of the tangent drawn from M, and consequently EM equals that tangent, which proves the Proposition.

3°. "Given a circle, and the length of the third diagonal of a quadrilateral inscribed in it, the distance from the centre to the middle point of this diagonal is determined."

In the demonstration of the last Proposition we found (see last figure) $OM^2 = EM^2 + OT'^2$, which proves the Proposition.

78. *Given a circle, and the lengths of the three diagonals of a quadrilateral inscribed in it, to construct the quadrilateral.*

The solution of this problem depends on polar properties. In addition to the principles already laid down, the following Lemma is required:—

LEMMA 5.—Given two sides of a triangle, and the length of a line drawn from the vertex cutting the base in a given ratio, the triangle can be constructed.

Analysis.— Let ABC be the required triangle, and let its base AB be cut in D, so that AD : BD is given. Draw BK (see fig.) parallel to AC. Since AC : BK :: AD : BD, BK is known; also since CD : KD :: AD : BD, KD is known, and therefore CK also. We have therefore the three sides of the triangle CBK which determine that triangle, and consequently the triangle ABC.

We shall now return to the analysis of the original problem:—

Let O be the centre of the given circle, and M″, M′, M the middle points of the three diagonals of the required

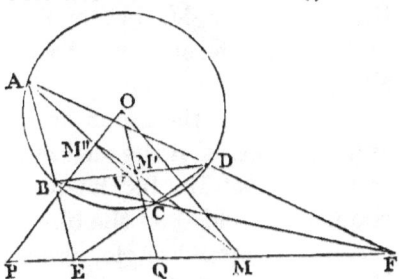

quadrilateral ABCD. Draw OM″ and OM′, and let the joining lines meet the third diagonal in P and Q respectively. Now, since the intersection of the diagonals AC, BD (which is marked V on the figure) is (Art. 41, 1°) the pole of the third diagonal, EF, it follows (Art. 40, 2°) that the pole of AC is on the line EF, and is therefore the point P; and for the same reason the pole of the line BD is the point Q. We have then (Art. 38) OP·OM″ = the square of the radius = a given quantity. But, since AC is given, and the radius also given, OM″ is evidently known; and, therefore OP is known. For a like reason OQ is known. Also, since M, M′, M″ are in one right line (Art. 71, 2°), we may consider MM″ as a transversal cutting the sides of the triangle OPQ, and therefore (Art. 9, Lemma 2) $\dfrac{PM}{MQ} = \dfrac{PM'' \cdot OM'}{QM' \cdot OM''}$ = a known quantity. We have then, finally, to construct a triangle OPQ, in which are given two sides OP, OQ, and the line OM (which is known by Prop. 3° of last Article) cutting the base in a given ratio. This can be done by means of the Lemma, and the construction of the triangle evidently determines the quadrilateral required.

79. *Given the base of a triangle in magnitude and position; given also the difference of its sides; it is required to prove that the polar of the vertex, with respect to a circle of given radius whose centre is at one extremity of the base, constantly touches a given circle.*

In order to prove this Proposition we shall require the following Lemma:—

LEMMA 6.—Given the base and the difference of the sides of a triangle, if a perpendicular be drawn from either extremity of the base on the line bisecting the vertical angle, its foot lies on the circumference of a given circle.

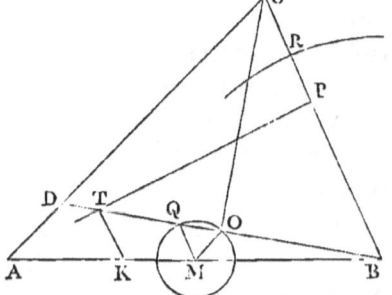

Let AB be the given base, and C the vertex of the variable triangle. Let BO be the perpendicular from B on the bisector of the vertical angle, and let it be produced to D (see fig.). Now, it is evident from the construction, that AD = the difference

ADDITIONAL EXAMPLES ON THE THEORY OF POLARS.   83

of the sides, and is therefore given. It is also evident that BD is bisected in O; and therefore the line joining O to M, the middle of the base, is parallel to AD, and equal to $\frac{1}{2} \cdot$ AD. The point O lies therefore on the circumference of a circle whose centre is the middle point of the base, and whose radius equals $\frac{1}{2} \cdot$ the given difference of the sides.

We shall now prove the theorem above stated.

Let BR be the radius of the given circle whose centre is B (see last figure), and PT the polar of the vertex C. Let BO, the perpendicular from B on the bisector of the vertical angle, cut the circle, on which O lies (by Lemma 6), again in Q, and let T be the point where the same perpendicular cuts the polar. Draw TK parallel to BC; then, a circle having K for centre, and TK for radius, will possess the properties asserted in the enunciation.

For, join M, Q. Then, the angle MQO = MOQ = CDB (since MO is parallel to AD) = CBD; therefore MQ is parallel to CB, and therefore to TK. Also (since C is the pole of PT), $BR^2 = CB \cdot BP = TB \cdot BO$ (since CB : BO :: TB : BP). The rectangle TB · BO is therefore a given quantity, and the rectangle BQ · BO also being (Euclid, B. iii. Prop. 36) a given quantity, the ratio of these rectangles, namely TB : BQ, is given. It follows that the ratios BK : BM and TK : QM are both given, and therefore that the point K is fixed, and the length of TK constant. Since the angle KTP is a right angle, a circle described with K as centre, and TK as radius, will touch the line PT at the point T, and the Proposition is proved.

80. LEMMA 7.—Given the base and the sum of the sides of a triangle, the foot of a perpendicular from either end of the base on the external bisector of the vertical angle lies in the circumference of a given circle.

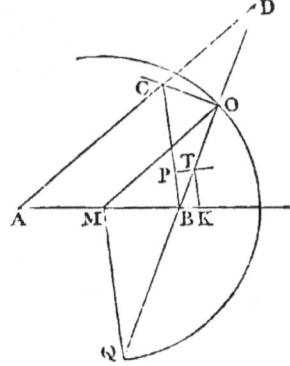

Let ABC be the triangle, and CO the external bisector of the vertical angle, and BO the perpendicular on it from B. It is evident from what has been said in proving the last Lemma, that MO is parallel to AD, and equal to ½ · the given sum of sides. The point O lies therefore on a circle having the middle point M of the base for centre, and half the sum of sides for radius.

This Lemma leads to a theorem analogous to that given in the last Article:—

Given the base and the sum of the sides of a triangle, the polar of its vertex with respect to a circle of given radius, whose centre is at one end of the base, constantly touches a given circle.

Let PT (see last figure) be the polar of the vertex C with respect to a circle whose centre is B, and whose radius is given. (This circle is not drawn on the figure). We have then BO · BT = BC · BP = a given quantity, and also BO · BQ = a given quantity (Euclid, B. iii. Prop. 35). The rest of the proof goes on exactly as before.

81. From the last two Articles the following theorem is derived:—

If a circle touch two given circles (the nature of the contacts being given), the polar of its centre with respect to one of the given circles touches a given circle.

For, if we consider the centre of the variable circle as the vertex of a triangle whose base is the line joining the centres of the given circles, it will be found that in all cases the sum or difference of the sides remains constant. This will be seen by the inspection of the figures of the various cases, two of which are annexed.

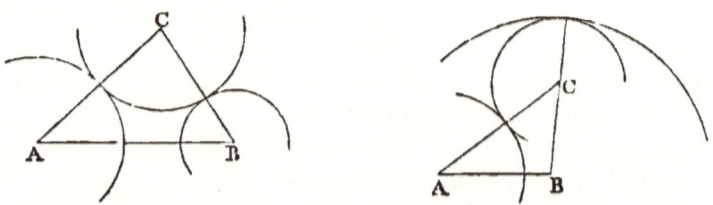

In both figures, A and B represent the centres of the given circles, and C the centre of the variable circle. In the first, AC − BC remains constant, and in the second AC + BC.

82. The middle point of the polar of a given point with

respect to a given circle (see Art. 48, Prop. 2°) possesses other properties worth noticing in addition to those contained in the Proposition referred to.

1°. "If lines be drawn from a given point and the middle of its polar to any point on the circle, their ratio is constant."

This is evident from the proof of Art. 39. For (see figs. of that Art.), since the triangles COQ and CQO' are similar, we have OQ : CO :: O'Q : CQ, and therefore OQ : O'Q :: CO : CQ.

2°. "If any chord be drawn through a given point, the rectangle under the distances from its extremities to the middle of the polar of the point is constant."

For (see the same figs.), when the pole is *within* the circle, we have, since the angle PO'Q is bisected, $PO' \cdot QO' = PO \cdot QO + OO'^2$ (see Lardner's Euclid, B. vi. Prop. 17). Now, as the circle and the pole O are supposed to be given, the rectangle $PO \cdot QO$ is constant (Euclid, B. iii. Prop. 35), and the Proposition is proved.

When the pole O is *without* the circle, we have the angle PO'Q bisected externally, and (Lardner's Euclid, B. vi. Prop. 17) $PO' \cdot QO' = PO \cdot QO - OO'^2 =$ a constant quantity (Euclid, B. iii. Prop. 36).

We might have proved both cases of this Proposition more concisely (Euclid, B. iii. Props. 36 and 35) by observing that the segment O'Q is equal to the distance from O' to the second point in which PO' cuts the circle.

83. The Proposition just proved leads to the following general theorem :—

1°. *If a triangle be inscribed in a given circle so that one side may pass through a given point, the remaining sides intercept on the polar of the point segments, which (measured from the middle of the polar) contain a constant rectangle.*

Let PQR be the triangle inscribed in the given circle (see fig.), O the given point, and O' the middle of its polar. Join PO', QO', PQ'. Now, since the angle PO'Q is bisected, O'P = O'Q', and the line PQ' is therefore perpendicular to the diameter, and consequently parallel to the polar ST. We have then the angle PTO' = RPQ' =

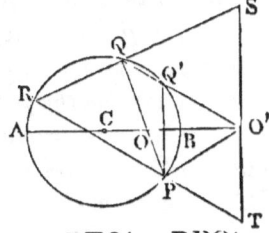

(Euclid, B. iii. Prop. 22) SQO', and therefore (the angles SO'Q and PO'T being equal) the triangles SO'Q and PO'T are similar, and SO' : O'Q :: PO' : O'T; whence SO' · O'T = O'Q · PO', a constant quantity (Art. 82, Prop. 2°). The Proposition enunciated is therefore proved when the pole is within the circle, and a similar proof applies when it is without.

By taking the point R at either end of the diameter AB, we may deduce the following result:—

2°. *If a chord be drawn through a given point in the diameter of a given circle, the lines joining its extremities to one end of the diameter intercept, on a given perpendicular to the latter, segments, which contain a constant rectangle.*

Let KS' be the given perpendicular to the diameter, O and O' remaining as before; we have to prove that when the chord PQ revolves round the given point O, the rectangle KS' · KT' is constant.

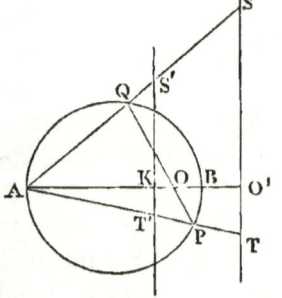

By similar triangles we have $\dfrac{KS'}{O'S} = \dfrac{AK}{AO'}$, and $\dfrac{KT'}{TO'} = \dfrac{AK}{AO'}$; therefore, by compounding the ratios, $\dfrac{KS' \cdot KT'}{SO' \cdot TO'} = \dfrac{AK^2}{AO'^2}$, that is, KS' · KT' : SO' · TO' is a given ratio; but SO' · TO' is constant by the theorem above proved, and therefore KS' · KT' is constant.

In the Proposition just proved, let the point K coincide with B, the other end of the diameter, and we shall have another theorem:—

3°. *If through a given point on a diameter a chord be drawn cutting a given circle, tangents at its extremities will intercept, upon a tangent at one extremity of the diameter, segments which, measured from the point of contact, contain a constant rectangle.*

We have to prove (see figure on next page) that when the chord PQ revolves round the given point O, the rectangle BM · BN remains constant.

Join BQ, BP. Then, since MQ and MB are tangents, they

are equal, and (BQS' being a right angle) M is the middle point of BS', and in like manner N is the middle point of BT'. It follows that the rectangle BM · BN is one-fourth of the rectangle BS' · BT', and is therefore constant, since the latter rectangle is so.

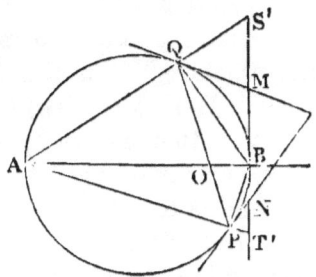

Hence, conversely, *if two tangents intercept upon a fixed tangent to a given circle segments containing a constant rectangle, the chord of contact passes through a fixed point.*

84. We shall conclude this part of our subject with the following problems:—

1°. *From a given point O in the produced diameter of a given circle, to draw a secant OPQ such that perpendiculars on the diameter from the points of section P and Q may intercept a portion ST of given length.*

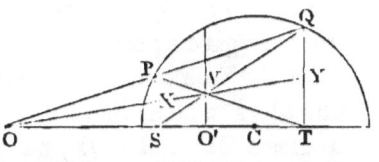

Analysis.—Join P, T, and Q, S, and through the intersection of the joining lines draw a perpendicular to the diameter. Now, the line OY (see fig.) is (Art. 9, 1°) cut harmonically at X and V, and therefore OQ is cut harmonically; consequently the perpendicular VO' is the polar of O, and therefore given in position. Of three lines in harmonic proportion, we have then the mean OO', and the difference of the extremes OS and OT, from which the extremes themselves can be found (Art. 69, 1°), and therefore the position of the secant determined.

2°. Retaining the other conditions of the last question, *let the ratio of* PS : QT *be given (the point O being no longer given)*; it is required to find the position of the line PQ.

Since PS : QT (see last figure) is a given ratio, OS : OT is a given ratio, and therefore, ST being also a given length, it is easy (Lardner's Euclid, B. vi. Prop. 10) to find the lengths of OS and OT. Again, since (Art. 9, 1°) OT is cut harmonically in S and O', SO' : O'T :: OS : OT, that is, in a given ratio, and therefore SO' and O'T can be separately determined. Finally, calling C the centre of the circle, and considering CO and CO′ as two unknown lines, we have their difference OO′ (which is equal to OS + SO′) and the rectangle under them (which is equal to the square of the radius, since VO′ is the polar of O), and therefore (Lardner's Euclid, B. ii. Prop. 10) the lines can be found. The position of PQ is then evidently determined.

Our next problem requires the following Lemma:—

LEMMA 8.—" Given a triangle ABC and a point O in the base produced, if a line OXY be drawn cutting the sides in such a manner that parallels XD, YD to the sides shall meet on the base, the line XY is the base of the *maximum* amongst all triangles inscribed in the given triangle, so as to have their bases passing through O, and their vertices at another given point E, also on the base."

Let PQE be one of the inscribed triangles, and conceive perpendiculars to be drawn from D and E to the line OPQ; then, the triangles PQD and PQE, having a common base PQ, are to one another as the perpendiculars, that is, as OD : OE, which is a given ratio. It follows that the triangle PQE is a maximum when PQD is a maximum. Now the latter triangle is a maximum when PQ coincides with XY; for, joining PY, we have the triangle DXY equal to DPY, which is evidently greater than DPQ, because QY is parallel to DX. This proves the enunciation above given.

3°. *In a given circle it is required to inscribe a triangle so that its vertex shall be at a given point, that its base shall pass through another given point on the tangent drawn at the former point, and that its area shall be a maximum.*

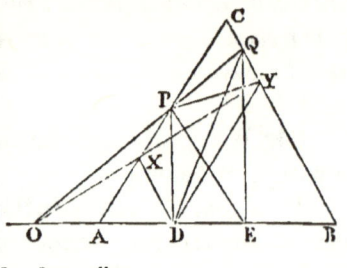

Let E be the given vertex, and O the given point on the tangent at E. Now it is easy to prove, in the first place, that if OY could be drawn so that parallels XD, YD

ADDITIONAL EXAMPLES ON THE THEORY OF POLARS.

to the tangents at the points Y, X should meet on the given tangent OE, the triangle XYE would be the triangle required. For, it would be a maximum (Lemma 8) amongst all triangles inscribed in the *triangle* ABC (see figure), having their vertex at E, and their bases passing through O, which proves *(a fortiori)* that it would be greater than any other triangle inscribed in the *circle* under the prescribed conditions.

This being understood, we shall now continue the analysis. Since C is the pole of XY, and E the pole of OE, the line joining C and E is the polar (Art. 40, 1°) of O, and therefore passes through the point of contact T of the second tangent drawn from O. Again, since CE is cut harmonically, the lines OC, OT, OY, and OE make an harmonic pencil, and therefore cut the line CD harmonically in V and F. We have then CV : VF :: CD : FD, or :: 2 : 1 (since CXDY is a parallelogram); and therefore DF (which is equal to CF) = 3 · VF. Now the angle OFD is a right angle, since the tangents CX, CY are equal, and therefore the determination of the line OY finally comes to this:—" Given the vertical angle of a triangle (namely the angle DOV) and the ratio in which its base (DV) is divided internally by a perpendicular from the vertex, to find the angles made by the perpendicular with the sides." To complete the solution, take any right line and divide it into two segments, one of which equals three times the other; at the point of section raise a perpendicular to meet the circumference of a segment of a circle described on the divided line, and containing an angle equal to the given vertical angle, and join the point of intersection of the perpendicular and circle to the ends of the divided line. The angle subtended by the greater segments is equal to that which the required line OY makes with the given line OE.*

---

* Since $\frac{DF}{FO} = \tan DOF$, and $\frac{VF}{FO} = \tan VOF$, the required line OY divides the given angle TOE in such a manner that $\frac{\tan DOF}{\tan TOF} = 3$. The angles may be computed separately as follows:—

$$\frac{\tan DOF + \tan TOF}{\tan DOF - \tan TOF} = \frac{4}{2} = 2 \; ; \text{ therefore } \frac{\sin (DOF + TOF)}{\sin (DOF - TOF)} = 2, \text{ and } \sin (DOF - TOF) = \frac{1}{2} \cdot \sin TOE.$$ We have then the sum and difference of two angles, and consequently the angles themselves.

When the line OE is equal to the radius of the circle, the angle TOF = 90°, and

ADDITIONAL EXAMPLES ON RADICAL AXES AND CENTRES OF SIMILITUDE.

85. We shall conclude this Chapter with a few additional Propositions concerning *radical axes and centres of similitude*.

1°. *If a system of circles have a common radical axis, to find the locus of the middle points of the polars* (see Art. 48, 2°) *of a given point, with respect to the circles of the system.*

Referring to the figure of Art. 56, it is evident that (taking R for the given point) the locus of all such points as V is the circle passing through R and cutting the given system orthogonally. The mode of describing this circle is explained in Art. 57.

2°. When the common chord of the system is *ideal*, the line joining the given point to the fixed point through which its polars pass (see Art. 57) subtends a right angle at either of the *limiting points*.

This is evident from Art. 55, 3°.

3°. Retaining the supposition of the last proposition, a common tangent to any two circles of the system subtends a right angle at either of the limiting points.

For, a circle on the common tangent as diameter cuts the two circles orthogonally, and therefore (Art. 55, 3°) passes through the limiting points.

86. "Given a circle TPT' and a point O (see fig.), let any two secants OQ, OQ' be drawn; it is required to find the locus of the point X, in which circles round the triangles OPQ', OP'Q will cut one another."

The three chords OX, PQ', P'Q meet in one point (Art. 62), lying (Art. 40, 3°) on the line TT', the polar of O. Again, the rectangle $TV \cdot VT' = PV \cdot VQ' = OV \cdot VX$, and therefore, a circle described through the three given points O, T, T', must pass through X, and is the locus required.

---

therefore tan TOF = cot DOF; therefore $\tan^2$ DOF = 3, whence tan DOF = $\sqrt{3}$, and DOF = 60°.

87. If two secants CV', CP' be drawn from either centre of similitude of two circles (suppose C), we have the following properties:—

1°. "A chord of one of the circles joining any pair of the points in which the secants cut the circle (such as OP), is parallel to the corresponding chord (O'P') of the other circle."

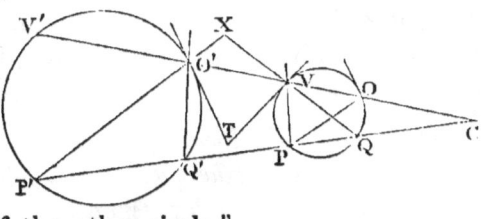

For (Art. 65, 2°), CO : CO' :: CP : CP', and therefore OP is parallel to O'P'; and a similar proof applies to the other cases.

2°. Hence, *tangents at two corresponding points are parallel*. This appears by considering them as indicating the limiting directions of evanescent chords. This property has been already stated in Art. 65, 2°.

3°. "Two chords inversely corresponding (such as VQ, O'P) meet on the radical axis of the two circles."

For (Art. 65, 3°), CV · CO' = CQ · CP'; therefore the quadrilateral O'P'QV is inscribable in a circle; and therefore XO' · XP' = XV · XQ. Tangents drawn from X to the two circles are therefore equal, and the Proposition is proved.

4°. It follows from the last that *tangents drawn at two points inversely corresponding* (such as O' and V) *intersect on the radical axis*. For, if we conceive the secant CP' to revolve round C until P coincides with V, and Q' with O', the chords VP and O'Q' constantly meet on the radical axis, and will continue to do so in their limiting state.

88. *If four secants be drawn from either centre of similitude of two circles, the anharmonic ratio of any four of the points where the secants cut one of the circles is the same as that of the four points of the other circle corresponding, all of them directly, or else all inversely, to the former.*

In order to prove this let us conceive two other secants, such as CV' (see last figure), to be drawn. We shall thus have two other points such as O, which we may represent by O_1 and O_2, and also another pair such as O', which we may call O'_1, O'_2. Let us now conceive P to be joined to O_1 and O_2, and P′ to be joined to O'_1 and O'_2, and it will appear at once (Art. 87, 1°) that the anharmonic ratio $P \cdot QOO_1O_2 = P' \cdot Q'O'O'_1O'_2$. This proves the Proposition in the case of points directly corresponding, and that of points inversely corresponding follows from the former, by Art. 40, 4°.

89. *To find the locus of the point of contact of two circles which touch one another, and each of which also touches two given circles, either both externally or both internally.*

It follows from what has been said in Art. 66, 2°, that the common tangent drawn at the point of contact of the two variable circles passes through the external centre of similitude of the two given circles, and therefore that the locus required is a circle with that point for centre, and the tangent as radius.

90. *Required the "envelope" of a system of circles touching a given circle, and cutting another given circle orthogonally.*

(The "envelope" of a system of curves or right lines means a fixed *line* which is touched by each member of the system.)

Let X be the centre of the variable circle which touches at O the given circle, whose centre is A, and cuts orthogonally the circle whose centre is C, and radius CT. Join CO, and produce the joining line to V (see fig.); join also CA, XA, and XV′, 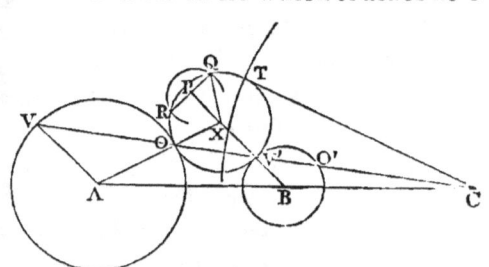 and let this last drawn line cut CA in B. We have then, by the conditions of the question, $CV \cdot CO =$ a constant, and $CV' \cdot CO = CT^2$ a constant also, and therefore $CV : CV'$ is a constant ratio. It follows (AV and BV′ being parallel) that B is a fixed point, and BV′ a constant length; and therefore a circle described with B as centre, and BV′ as radius, is the envelope sought.

This Proposition is given in the Lady's Diary for 1851. It is evidently the converse of Prop. 2° of Art. 66.

91. *Given the hypotenuses of two right-angled triangles, and*

the sum of one pair of their sides, it is required to construct them so that the sum of the other pair shall be a maximum.

Let AB be the given sum of the pair of sides of the two required right-angled triangles. With A and B as centres, describe circles having for radii the given hypotenuses. Through their internal centre of similitude, C', draw EF, perpendicular to AB; then, it is evident from Prop. 1° of Art. 66, that AEC' and BFC' are the two right-angled triangles required.

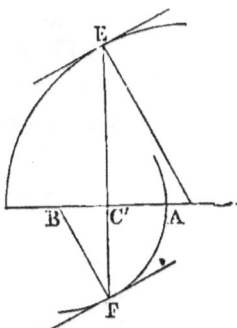

92. LEMMA 9.—If AT and TN (see fig.) be tangents to a circle, and NX a line inflected on the diameter, and equal to TN, and if from N two other lines be drawn, one, NP, perpendicular to the diameter, and the other, NO, bisecting the angle TNX, a right line, OY, drawn from O perpendicular to NP, is equal to the radius of the circle.

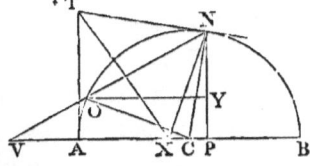

Produce NO to V, and draw the radii NC, OC. From the conditions of the question it is evident that NV is perpendicular to TX; and therefore the angle NOY (which is equal to NVX) is equal to the angle ATX. We have then, by similar triangles, OY : ON :: AT : TX, or :: TN : TX. Again, the angle TNX, being the double of TNO, is equal to (Euclid, B. iii. Props. 20 and 32) the angle NCO; therefore the triangles TNX and NCO, being isosceles, and having equal vertical angles, are similar, and consequently TN : TX :: NC : ON. It follows that OY : ON :: NC : ON, which proves that OY = NC.

The following problem depends on the Lemma above given:—

From a given point V *in the produced diameter,* AB, *of a given circle, to draw a secant* VN, *so that the inscribed quadrilateral* ABNO (see fig.) *shall be a maximum.*

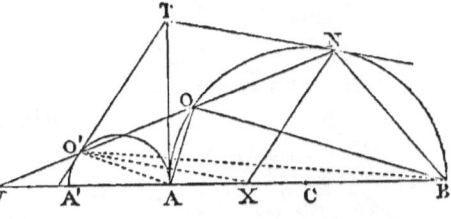

Analysis.—Join B, O; draw AO' parallel to the joining line, and join BO'. We have then a triangle, BNO', equal to the quadrilateral. Also, since VO' : VO :: VA : VB, it is evident that if VN be considered as a line revolving on V, the locus of the point O' is a circle having with the given circle the point V as their external centre of similitude. This circle is determined by taking VA' : VA :: VA : VB, which proportion gives the position of A' one extremity of its diameter, the given point A being the other. Again, the points O' and N being inversely corresponding, tangents drawn at these points meet (Art. 87, 4°) on the radical axis of the two circles, that is, on their common tangent AT. Now it is easy to see that the triangle BNO' is (Art. 84, Lemma 8) a maximum when VN is drawn so that O'X and NX, parallel to the tangents at N and O', intersect on the line VB; in which case the parallelogram TNXO' is evidently a rhombus, and the angle TNX bisected. It appears then by the foregoing Lemma that the position of the line VN, which solves the problem, is such that perpendiculars from O and N, on the diameter AB, intercept a portion equal to the radius CA. The question is therefore finally reduced to Prop. 1° of Art. 84.

93. The succeeding problems depend on the principle contained in Art. 65, 4°.

1°. *To describe a circle touching two given circles, and bisecting the circumference of a third.* (The nature of the contacts is supposed to be given.)

Let A and B be the centres of the given circles which the required circle whose centre is X (see fig. in Art. 90) is to touch and let P be the centre of the third given circle whose circumference is to be bisected. Now, by Art. 65, 4°, the line joining the points of contact O and V' passes through one of the centres of similitude C, and, those points being points inversely corresponding, the rectangle CO · CV' is (Art. 65, 3°) given, and therefore (Lardner's Euclid, B. ii. Prop. 6) $CX^2 - OX^2$ is a given quantity. Again, since, by the conditions of the question, the points Q, P, R lie in a right line, the angle QPX is a right angle, and $QX^2 - PX^2$ is a given quantity. It follows (OX and QX being equal) that $CX^2 - PX^2$ is a given quantity, and therefore (Note on Art. 58) that X lies on a known right line. The question comes then to this:—" To describe a circle touching two given circles, and having its centre on a given right line," the solution of which will appear from the Note on Art. 66, 3°.

The figure represents the case when the line joining O and V′ passes through the external centre of similitude, but a similar analysis applies in the case of the internal centre.

2°. *To describe a circle touching two given circles, and cutting a third orthogonally.* (The nature of the contacts being given as before.)

The investigation of this problem is so like the last, that we shall leave it to the student, merely observing that the angle PQX is now a right angle in place of QPX. (See the figure in Art. 90.)

94. We shall place here some additional theorems connected with a system of three circles.

1°. *In a system of three circles the lines joining each of the three centres to the internal centre of similitude of the other two circles meet in a point.*

For (see Art. 67), we have $\frac{AC''}{BC''} = \frac{R}{R'}$, $\frac{BA''}{CA''} = \frac{R'}{R''}$, and $\frac{CB''}{AB''} = \frac{R''}{R}$; therefore, by compounding the ratios, $\frac{AC'' \cdot BA'' \cdot CB''}{BC'' \cdot CA'' \cdot AB''} = \frac{R \cdot R' \cdot R''}{R' \cdot R'' \cdot R} = 1$. The Proposition follows by Lemma 1 of Art. 9.

2°. We saw in Art. 68 that four pairs of circles can be described to touch three given circles, each pair having the same internal centre of similitude, namely, the radical centre of the three given circles, and having for its radical axis a corresponding axis of similitude.

From these properties it immediately follows that *a perpendicular drawn from the radical centre to the axis of similitude corresponding to a pair of touching circles passes through their centres.*

3°. It appears from Art. 54 that each of the pairs of circles above mentioned may be considered as determining an infinite system of circles having a common radical axis. We shall now prove that *a circle orthogonal to the three given circles belongs to each of the four systems thus determined.*

For, a circle orthogonal to the three circles must have for its centre the radical centre R (see fig. of Art. 68), and for its radius the tangent RV (Art. 55). Now since T is the pole of P′Q′ (which (Art. 68) passes through R), the chord VV′ of the orthogonal circle (being the polar of R) must (Art. 40, 1°) pass through T, and therefore the square of the tangent from T to that circle $= TV \cdot TV' = TP'^2 = TQ'^2$. This proves that the tangent from T to the orthogonal circle is equal to that from T to either circle of the pair in question. A similar result holds good for the other points analogous to T, and the Proposition immediately follows.

CHAPTER VI.

THE PRINCIPLE OF CONTINUITY.

95. We have referred, on two or three occasions, to *the principle of continuity*, and in this Chapter we propose to explain its elementary geometrical relations. The principle in question must have been to some extent employed in every period of geometrical science; but it is only in modern times that its importance as a mode of discovery or of proof has been adequately estimated. In the writings of Monge, and the members of his distinguished "school," are to be found many beautiful results of this principle in the higher departments of geometry; and it cannot be entirely useless to illustrate, by simpler examples, its meaning and validity, as well as the method of applying it. The following may be taken as a statement of the principle itself:—

"Let a figure be conceived to undergo a certain *continuous* variation, and let some *general* property concerning it be granted as true, so long as the variation is confined *within* certain limits; then the same property will belong to *all* the successive states of the figure (that is, all which admit of the property being expressed), the enunciation being *modified* (occasionally) according to known rules."

96. In order to elucidate the foregoing Proposition, we shall commence with a class of cases which very commonly occur,

namely, where, having proved some general property of a figure, we wish to apply it in an extreme or *limiting state*, as it is called. In these cases the principle of continuity is identical with the fundamental principle of the *method of limits*, which, when considered geometrically, assumes that whatever is true of a variable figure *up to* a certain limit is also true *(mutatis mutandis)* at the limit. The following are examples:—

1°. Suppose it proved (Lardner's Euclid, B. ii. Prop. 6) that when any right line (such as OQ) is drawn from a given point O, so as to *cut* a given circle whose centre is C, the rectangle $QO \cdot PO = CO^2 - CP^2$ (O being supposed to be without the circle). What does this property become when the right line OQ, being conceived to revolve round the point O, ceases to *cut* the circle, and becomes a tangent? In this limiting case we have, of course, $OT^2 = OC^2 - CT^2$.

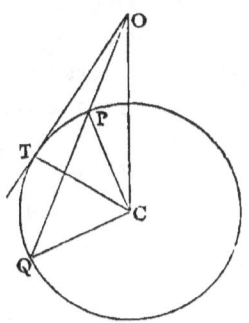

2°. Again (see last figure) the angle $CPQ = CQP$. What does this property become in the same extreme case? Since the three angles of the triangle CPQ, are equal to two right angles, and the angle PCQ vanishes when the points P and Q coincide, the result is that the limiting value of CQP, which is CTO, is a right angle.

3°. The opposite angles of a quadrilateral inscribed in a circle are supplemental. From this we infer, in a similar manner, that the angle made by a tangent and a chord drawn from the point of contact equals the angle in the alternate segment. (Lardner's Euclid, B. iii. Prop. 32).

These examples are sufficient to show the mode of deducing properties of tangents to a circle from those of secants or chords. This process has been frequently employed in the foregoing pages. See Arts. 28, 46, 75 (Prop. 1°), 87 (Props. 2° and 4°).

4°. *When two tangents are drawn to a circle or any other curve, and their points of contact become coincident, their point of intersection reaches a limiting position, namely, the point of contact itself.* This Proposition may be taken as self-evident.

O

We may verify it in the case of the circle, by recollecting that the intersection of two tangents is the pole of the chord of contact. The principle in question has been employed in Arts. 42 (Props. 2° and 4°) and 47.

The principle of continuity, taken in the point of view presented in this Article, has also been made use of in Arts. 3, 4, 21, 24, 54, 55 (Prop. 3°), 70 (Prop. 1°), 73 (Prop. 1°).

97. We shall now give some examples in which the modifications of a property result from the changes in the relative position of certain parts of the figure produced by the successive variation mentioned in Art. 95. In order to understand these examples, the reader will require to be acquainted with the fundamental rule of the application of algebra to the geometry of position. (See Note to Art. 5.)

1°. It is proved in the fifth Proposition of the second Book of Euclid, that if AB be a given line, C its middle point, and D *any point between* A and B, $AD \cdot BD + CD^2 = CB^2$. What does this Proposition become when the point D, supposed to move along the line AB (considered as extending indefinitely in both directions) takes a position such as D'?

The principle of continuity assumes that the property expressed in the equation above given, being true for an infinite number of successive positions of the point D, is a general relation between the distances of a variable point of the line from the three given points A, C, B, and must continue to be true for any position whatever of the point *on the indefinite right line*, the expressions of the distances being modified according to the established rule above referred to. Now, when D passes into the position D', its distance from B changes its sign, and the equation becomes $AD' \cdot - BD' + CD'^2 = CB^2$, that is, $- AD' \cdot BD' + CD'^2 = CB^2$, and therefore $CD'^2 = CB^2 + AD' \cdot BD'$, which is the sixth Proposition of the second Book of Euclid.

2°. If ABC (see figure on next page) be an obtuse angled triangle, and CDB a right angle, we have (Euclid, B. ii. Prop. 12) $AC^2 = AB^2 + BC^2 + 2AB \cdot BD$. Let us suppose the angle

ABC gradually to become less, so as to pass into an acute angle. What does this equation then become? In this case it is evident that the distance BD will change its sign, and we shall have the thirteenth Proposition of the second Book of Euclid.

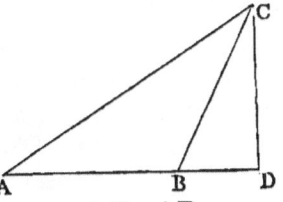

3°. Let CD (see fig.) bisect the angle ACB, and we have (Euclid, B. vi. Prop. 3) $\frac{AC}{CB}=\frac{AD}{DB}$. Now let the right line AB revolve round the point D, so as to come into the position A'B'. How must the equation be modified? In this case the distance CA evidently changes its sign. Also DA changes its sign relatively to DB, we have, therefore,

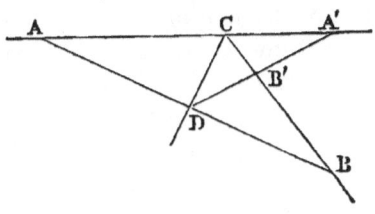

$\frac{-CA'}{CB'} = \frac{-A'D}{B'D}$, or $CA' : CB' :: A'D : B'D$.

4°. Again (Lardner's Euclid, B. vi. Prop. 17) granting that (see last figure) $CD^2 = AC \cdot CB - AD \cdot DB$, we infer by the principle of continuity that $CD^2 = -A'C \cdot B'C + A'D \cdot B'D$, or $CD^2 + A'C \cdot B'C = A'D \cdot B'D$.

5°. When two chords are drawn through any point O *within* a circle, we have $OP \cdot OQ = OR \cdot OS$. Now let O move gradually until it gets without the circle. It is manifest that OP and OR change their signs relatively to OQ and OS, and the equation becomes $-O'P' \cdot O'Q' = -O'R' \cdot O'S'$, and therefore $O'P' \cdot O'Q' = O'R' \cdot O'S'$.

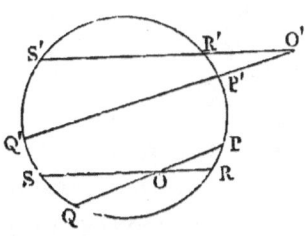

In the class of cases which we are now considering, the principle of continuity is identical with that laid down by Carnot, under the title "la corrélation des figures" (see his Geometry of Position). He proceeds, however, in a manner somewhat different to obtain the requisite modifications in his "correlative figures." For instance, his mode of derivation in the last example is as follows:—

The equation $OP \cdot OQ = OR \cdot OS$ may be written $OQ \cdot (PQ - OQ) = OS \cdot (RS - OS)$. Now when O comes into the position O', we shall still have $O'Q' (P'Q' - O'Q') = O'S' \cdot (R'S' - O'S')$ (since all the quantities entering into the former equation remain finite during the transition), and, changing the signs on both sides, we find $O'Q' \cdot (O'Q' - P'Q') = O'S' \cdot (O'S' - R'S')$; that is, $O'Q' \cdot O'P' = O'S' \cdot O'R'$, as before. (Géométrie de Position, p. 47).

98. In the last Article the properties under consideration involved the *magnitudes* of certain quantities. There are others of a different kind which we may mention here. Such are those which refer to certain points as lying on the same right line, or to certain lines as meeting in a point. It is in general sufficient, in these cases, to affix the same letters constantly to the points of intersection of coresponding lines (right lines being always considered as infinite in both directions) in order to recognise the varieties of the original Proposition resulting from a change of figure.

The examples given in Arts. 18, 19, 29, 33, will serve as illustrations of what has been said.

99. We come now to a class of cases which deserves particular attention as being that in which the power of the principle under consideration has appeared to greatest advantage. The cases referred to are those in which the variation of the figure is such that some parts of it cease to exist, or (speaking algebraically) become *imaginary*.

1°. The proofs of Propositions 1° and 2° of Art. 40 depend on this, that any right line through a point taken as pole, is cut harmonically by the circle and the polar. We saw, however, that in a certain case one pair of the harmonic conjugates become imaginary. Now the principle of continuity implies that the result is not at all affected by circumstances of this nature, (*provided that the enunciation of a Proposition do not cease to have a geometrical meaning* from its involving directly or indirectly the parts of the figure which have become imaginary). If, for example (see the second figure of Art. 39), the first of the two theorems in question is known to be true so long as R is *anywhere* on the portion of the given line *within* the circle, we conclude that this is a general property of the indefinite line, and will continue to be true when R passes *without* the circumference. This conclusion is verified, accordingly, by the proof given in the Note to the Article referred to.

2°. The two Propositions of which we have been speaking are true, no matter how small the square of the radius of the circle may be. They should therefore be true even when the square of the radius becomes negative, that is, *when the radius is imaginary*, provided that the enunciation still retain a geometrical meaning. Let us examine this case, supposing that the centre of the circle remains a real point.

From the definitions of pole and polar, with respect to a given circle (see Art. 38), it appears that their distances, measured from the centre, are such that their *algebraical* product equals the square of the radius. If, then, the square of the radius be negative, one of the distances must become negative, which indicates that the pole and polar must lie on *different* sides of the centre. 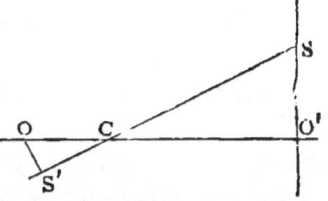 Let C (see figure) be the centre of the imaginary circle, and $-K^2$ the negative value of the square of the radius; and let CO′ be perpendicular to SO′. Take on the production of O′C a portion CO, such that $CO \cdot CO' = K^2$, then O will be the pole of SO′. This being understood, let S be any point on the polar, and OS′ a line through the pole perpendicular to CS produced through C, and it immediately appears that $CS \cdot CS' = CO \cdot CO' = K^2$, and consequently that the point S and the line OS′ (being at different sides of C) are pole and polar in the present point of view. This proves that Prop. 1° of Art. 40 holds good, and it is evident that the same is true of Prop. 2°. The results indicated by the principle of continuity, are, therefore, completely verified in the present case.

3°. We proved in Prop. 3° of Art. 37 that a transversal cutting a circle and the sides of a quadrilateral inscribed is cut in involution. Again, we proved (Art. 34) that when five out of six points in involution are given, two definite pairs being conjugate, the sixth point is determined. Hence it follows that if a quadrilateral be inscribed in a given circle, so that three of its sides pass respectively through three given points in a right line *which cuts the circle*, the fourth side also constantly passes through a fixed point in the same right line. Let us now suppose that the line containing the three given points moves gradually until it no longer meets the circle, and we infer by the principle of continuity that the result before arrived at still holds good, even though the former method of proof does not now admit of being applied. Accordingly, the reader will find this conclusion verified in Art. 49, where we have given a demonstration of an entirely different kind.

4°. Our next example requires the following theorem:—

If three circles be drawn through the same two points, tangents drawn to two of them from any point of the third are in a constant ratio.

Let E be the variable point from which the tangents are drawn. Join EC, EB, AB, and BD (see fig.). It is evident that as the line AE revolves round C, the angle BAC remains constant (Euclid, B. iii. Prop. 21). For a like reason the angles BDC and BEC remain constant; and, therefore, the triangles BAD and BAE are given in species. The ratio $\frac{AE}{AB}$ is therefore constant, and also $\frac{AB}{AD}$, and consequently the ratio compounded of these two, namely, $\frac{AE}{AD}$. But the ratio $\frac{AE}{AD}$ being fixed, the ratio AE : DE is fixed, and therefore the ratio of the rectangle AE · CE : DE · CE, or the ratio of the squares of the tangents drawn from E. It follows that the ratio of the tangents themselves remains constant. Q. E. D.

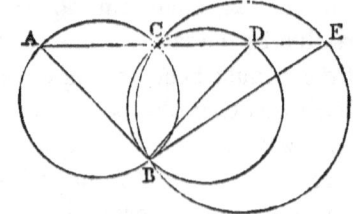

Let us now suppose the common chord of the three circles to become *ideal* (see Art. 53), and we shall have the following result:

If three circles belong to a system having a common radical axis (see Art. 54), *tangents drawn to two of them from any point on the third are in a constant ratio.*

100. It is easy to infer conversely, from the last example, that *the locus of the intersection of tangents to two given circles having a given ratio is* (excepting the case of equality) *a circle belonging to the same radical system as the given circles.*

In the particular case where the ratio of the tangents is the same as that of the radii, the locus passes through the centres of similitude of the given circles, and has hence been called by Mr. Davies their *circle of similitude* (see the Lady's Diary for 1851.) This circle possesses the following property:—

"If from any point on the circle of similitude of two circles two tangents be drawn to each of the circles, the angles contained are equal."

The reader will find no difficulty in proving this.

In the class of cases which we have last given, the principle of continuity is denominated by Chasles the principle of *contingent relations*. By this expression he means to intimate what we have already declared (Art. 99, 1°),—that the change from real to imaginary of certain parts of a figure does not (under the conditions before stated) invalidate a result previously obtained.

101. In the foregoing pages we have attempted to give an idea of the meaning and mode of application of the principle of continuity considered geometrically. If we have not fully succeeded, it may be urged as an excuse that the applications of the principle are extensive, and have not hitherto been reduced into a system. Nor have geometricians as yet agreed with respect to the exact ground on which its validity is to rest. Poncelet seems to consider the principle as being in some sort axiomatic, and founded essentially on the continuous nature of the geometrical magnitudes, about whose properties it is conversant. Chasles, on the other hand, is of opinion (as to one class of cases at least), that our confidence in its infallibility is to be ultimately referred to that which we are accustomed to repose in the general processes of algebra.

Admitting the general correctness of Poncelet's view, we still think that there are cases in which it is satisfactory, if not absolutely necessary, to justify the use of the principle in question by reasoning of the following kind:—A certain property of a figure existing with a certain generality (see Art. 95), is known to be true (geometrically). Now this property being true, we may *conceive* an algebraical proof to be made out for it. But this proof must, from the nature of algebra, apply equally to all cases possessing an equal generality. We have a right, therefore, to extend the original property to all cases having that equal generality, (modifying the enunciation, if necessary, according to the change of figure).

102. We shall illustrate the preceding reasoning by applying it to a few of the examples already given.

1°. In the first example of Art. 97, the figure consists of an indefinite right line, with three fixed points on it, and another point D also on the line in a position of a certain generality, that is, anywhere between the fixed points A and B. Now concerning this figure, we are granted that $AD \cdot BD + CD^2 = BC^2$, which expressed algebraically, gives the *identical* equation $(a + x)(a - x) + x^2 = a^2$, where a represents the line AC, and x the portion CD. This equation being identical is entirely independent of the relative magnitudes of a and x, and therefore applies to points *beyond* A or B, as well as to points *between* them. For a point such as D′ we have therefore $(AC + CD')(AC - CD') + CD'^2 = AC^2$, that is, $AD' \cdot - BD' + CD'^2 = AC^2$, or $AD' \cdot BD' + AC^2 = CD'^2$, as before.

2°. In the third example of Art. 99, the figure consists of *any* quadrilateral inscribed in a given circle, so as to have three sides passing through three given points in a right line *which cuts* the circle. We are supposed to know (by means of the points where the right line cuts the circle) that the fourth side constantly passes through a fixed point in the same right line. Now if we were to prove this theorem algebraically (suppose by the method of co-ordinates), the only data we should make use of are the co-ordinates of the centre, those of the three given points expressed by general symbols, the length of the radius, and the algebraical relation among the co-ordinates of the points, expressing that they are *in directum*. The points of intersection of the line containing the points with the given circle would not enter at all into the proof, and therefore their being real or imaginary could have no effect on the result. We conclude from this that the original Proposition continues to be true *when the right line no longer meets the circle*, although the former mode of proof will certainly not continue to apply.

3°. In example 4° of Art. 99, the original figure consists of any three circles having a common chord and two tangents drawn to two of them from any point on the third. In the modified figure the common chord is ideal. Now since the common chord of two circles is a right line perpendicular to the line joining their centres, and cutting it into two segments, the difference of whose squares equals the difference of the squares of the corresponding radii, it follows that the algebraical mode of expressing the data in the original state of the theorem referred to must be exactly the same as that relating to the modified state. The circumstance of the intersection of the circles being real or imaginary, will therefore not enter into the algebraical proof of the original Proposition, and the extension of the latter in the example cited is consequently justified.

103. It may perhaps be objected that in certain cases the employment of the principle under consideration is nothing more than the usual application of algebra to geometry. It is undoubtedly true that its successful use, when taken in its widest extent, requires a familiar acquaintance with the ordinary processes of algebraic geometry. Nevertheless, even in those cases where we appeal to the aid of algebraic notions in employing the principle of continuity, the primary geometrical conception is never superseded. We borrow, in fact, the comprehensiveness of analysis *without the necessity for analytical calculations*, and thus invest the science of pure space with a power and facility unknown in the earlier stages of its history.

The peculiar advantage of the principle of continuity in geometry, whether considered as a method of discovery or of demonstration, consists in this, that *amongst all cases of equal generality we may select that one which presents the circumstances most convenient for establishing geometrical relations.* These relations, once proved, become immediately transferable (with or without modification) to the remaining cases; and in this way many of the results of Monge and Poncelet and Chasles have actually been obtained.

104. It follows from the remark made at the commencement of Art. 96, that the method of *limiting ratios* in geometry, or as it is generally called, the method of *infinitesimals*, is fundamentally connected with the principle of continuity. We shall here give a specimen of its mode of application in problems relating to maxima and minima, premising that *a quantity is said to be infinitely small when its limiting ratio to a finite quantity equals zero.*

Through a given point O (*see fig.*) *within the legs of a given angle,* ACB, *to draw a right line, so that the part intercepted by the legs shall be a minimum.*

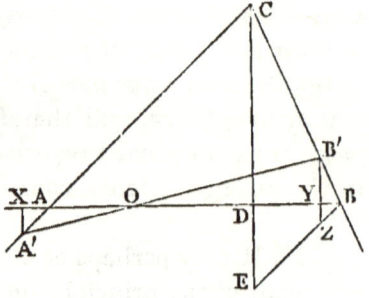

Analysis.—If we conceive a line drawn through O parallel to AC to revolve round O, so as to come into a position such as AB, and to continue to revolve in the same direction until it becomes parallel to CB, it is evident that the part intercepted by the legs of the angle being at first infinite, then finite, and afterwards infinite again, must be a minimum in some of its positions. This happens when two *consecutive* or infinitely close intercepts are equal; since the intercept then ceases to diminish.

Let AB be the minimum intercept, and A′B′ another intercept infinitely close to it, and let A′X, B′Y, and CD be drawn perpendicular to AB. Draw BE parallel to CA, and let it meet CD and B′Y produced in E and Z respectively. Then, since B′Y is perpendicular to OB, and infinitely small, it may be considered as an arc of a circle, and therefore OB′ = OY, and in like manner OA′ = OX, and therefore A′B′ = XY. Now, since AB is supposed a minimum, we have AB = A′B′. It follows then that AB=XY, and therefore AX = BY, and consequently A′X

$= ZY$. We have also, $\frac{BD}{AD} = \frac{DE}{DC} = \frac{ZY}{B'Y}$, and $\frac{AO}{OB} = \frac{OX}{OY}$ (since AX and BY are infinitely small), $= \frac{A'X}{B'Y}$. We have then finally $\frac{BD}{AD} = \frac{AO}{BO}$, from which it appears that $AO = BD$; and this is the characteristic property of the line in the required position.

It is easy to see from this result that *the required line is bisected when the given point O lies on the bisector of the given angle.*

The solution of the general problem cannot be completed by the use of the right line and circle permitted in Euclid's Elements. It may be solved mechanically by the contrivance employed by Philo in finding two mean proportionals (see Lardner's Euclid, B. vi. Prop. 13). This will be seen by joining the points O and C, and on the joining line as diameter describing a circle.

It follows from what has been said that the portion of the ruler intercepted by the legs of the right angle in Philo's method, above referred to, is a minimum.

105. The investigation given in the last Article suggests an elementary demonstration of the result arrived at. We shall give this demonstration for the sake of our less experienced readers.

Let us suppose (see the preceding figure) AB to be drawn so that $AO = BD$, and let A'B' represent any other line drawn through O, we have to prove that A'B' is greater than AB. For, (keeping the same construction as before,) since $AO = DB$, $\frac{OX}{OY}$ is a greater ratio than $\frac{BD}{AD}$, and therefore greater than $\frac{DE}{DC}$, or than $\frac{ZY}{B'Y}$. Hence $\frac{A'X}{B'Y}$ is greater than $\frac{ZY}{B'Y}$, or, A'X is greater than ZY, and therefore (the triangles AXA' and BYZ being similar) AX is greater than BY, and consequently XY greater than AB. As A'B' is evidently greater than XY, it is still greater than AB. In the same manner the Proposition may be proved when the point A' lies between A and C.

106. The result of the last Article may be extended as follows:—

Given any curve POQ, *the portion* AB *of a tangent intercepted by two given right lines* CA, CB, *is a minimum, when the segment* OA, *between the point of contact and one of the given lines, equals the segment* DB *between the other given line and the foot of the perpendicular* CD, *drawn upon the tangent from the intersection of the given lines.* (The curve is supposed to be such that the part of it between the given lines is concave towards their intersection, and the tangent is supposed to be drawn so that the points O and D fall between A and B.)

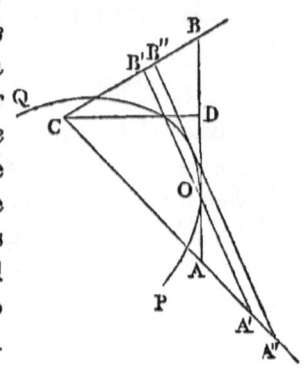

For, let A″B″ be the intercept of any other tangent, and let A′B′ be the intercept of a line drawn through O parallel to A″B″. We have, then, AB less than A′B′ by the last Article, and therefore less than A″B″.

If, for example, the curve be a circle, with C *for its centre, the tangent is a minimum when it is bisected by the point of contact.*

If the curve be a circle touching one of the given lines at their point of intersection, and having its centre on the other given line, the intercept of a tangent is a minimum when it is cut in extreme and mean ratio.

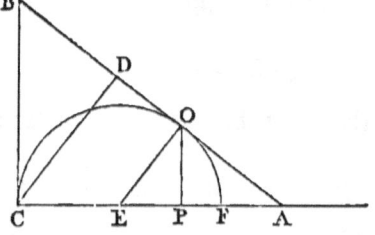

For, by what has been proved, the tangent AB is a minimum when BD = AO. Now, since ACB is a right-angled triangle, $AB \cdot BD = BC^2 = BO^2$; and therefore $AB \cdot AO = BO^2$. Q. E. D.

The point of contact O is found by cutting the radius EF of the given circle in extreme and mean ratio. Let P be the point of section, EP being the greater segment. A perpendicular to the diameter at the point P cuts the circle in the point required. We shall leave the proof to the reader.

CHAPTER VII.

ELEMENTARY PRINCIPLES OF PROJECTION.

107. In this Chapter we propose to explain some of the elementary principles of the *method of projection,* applications of which method will frequently occur in the remainder of this work.

"If right lines be drawn from any point *in space* to all the points of every right line or curve in a given figure, and if the entire system of right lines so drawn be made to intersect a surface *plane or curved,* a new figure is formed on the intersected surface, which is said to be the *projection* of the given figure." The point from which the right lines are drawn is called the *centre of projection,* and the intersected surface the *surface of projection.* From these definitions several properties of the projected figure immediately follow :—

1°. "Any two points on the projected figure subtend *the same angle* (or its supplement) at the centre of projection, as the *corresponding* points on the original figure." Hence it follows (by considering two coincident points) that *a tangent at any point of a curve is projected into a tangent at a corresponding point of the new curve.*

2°. "When the surface of projection is *a plane,* any right line in the original figure is projected into *a right line* in the new figure." Because the intersection of two planes is a right line.

3°. "When the surface of projection is *spherical,* with the centre of projection for its centre, a right line in the original figure is projected into *a great circle.*"

4°. "If several right lines meet in one point in the original figure, they are projected into right lines or great circles meeting in a corresponding point, according as the surface of projection is a plane, or a sphere, having the centre of projection for its centre."

5°. "If the given figure be considered as lying on a surface, it will be the projection of the new figure."

In this way, therefore, *each of the two figures is the projection of the other.*

When one of the surfaces is a sphere, we shall always suppose that its centre is the centre of projection, unless the contrary be distinctly stated; this being understood we may say that *the projection of a great circle on a plane is a right line.*

108. When the figure to be projected is a circle, the right lines drawn from the centre of projection to all the points in the circumference of the circle form a *cone*, which is said to be *right* when the line joining the centre of projection (that is, the vertex of the cone) to the centre of the circle is perpendicular to the plane of the circle. When the joining line is not perpendicular to the plane of the circle, the cone is said to be *oblique*. The circle we shall call the *base* of the cone.

The term cone may be taken in a wider sense than that in which we have defined it, but we shall in general use it in the sense above explained.

The lines drawn from the vertex to points on the circumference of the base are called *sides* of the cone.

It is evident that a lesser circle of the sphere may be regarded as the base of a right cone whose vertex is at the centre of the sphere. It is also evident that *a plane touching the sphere at the trigonometrical pole of the lesser circle will* (since the tangents of equal arcs are equal) *cut the cone in a second circle, which may be regarded as the projection of the former circle, and also as having the former for its projection.*

109. In questions of spherical geometry, the following principle of projection is sometimes of use:—

If two great circles be at right angles, and a plane be drawn touching the sphere at any point of one, the projections of the two circles on this plane will be also at right angles.

This is plain from the symmetry of the figure; but for the sake of illustrating some of the foregoing principles we shall give a demonstration:—

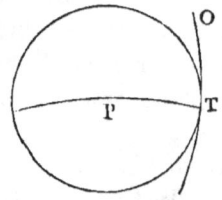

Let PT and TO be the great circles, P being the given point, and let us conceive a lesser circle to be described, with P as the trigonometrical pole, and the arc PT *as the spherical radius*.

This circle will touch the arc TO, as PTO is a right angle. Now, let us project the lesser circle and the great circles on the tangent plane, and we shall have a circle, a radius, and a tangent. (Arts. 108 and 107.) But the radius and tangent are at right angles; that is, the projections of the two great circles are at right angles. Q. E. D.

110. A very important principle, whether the surface of projection be plane or spherical, is as follows:—

"If the product of one set of right lines in a figure, divided by the product of another set in the same figure, equals a given number k, each line in the one set being *in directum* with a corresponding line in the other set, and the entire system of extremities of the lines in the one set being the same as the entire system of those of the other set; then,

"1°. The same equation holds good for any projection of the figure on *a plane*, substituting *the projection of each line in place of the line itself*.

"2°. The equation will be true for a *spherical* projection, provided that *the sines of the arcs into which the lines are projected* be taken *in place of the lines*."

In order to prove this Proposition, let AB be one of the right lines above mentioned, and let CP be the perpendicular drawn on it from the centre of projection C. Now, by plane trigonometry, $CA \cdot CB \cdot \sin ACB$ = twice the area of the triangle ACB = $AB \cdot CP$, and therefore

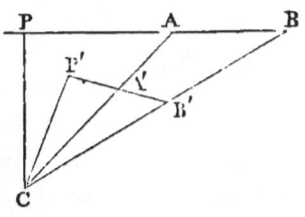

$AB = \dfrac{CA \cdot CB \cdot \sin ACB}{CP}$; and similar values will be found for the other lines entering into the given equation. Substituting these values in the equation, it will be seen, by attending to the data, that all distances such as CA, CB, CP will cancel one another, and there will remain an equation of the same form as the original, with the sines of the angles subtended at C by the various lines, such as AB, in place of those lines. (The *second* part of the Proposition is now evident, as the arcs are the measures of the angles at the centre.)

In order to prove the *first* part, let us conceive any plane projection of the original figure, and let $A'B'$ *correspond* to AB. Then $\sin ACB = \sin A'CB' = \dfrac{A'B' \cdot CP'}{CA' \cdot CB'}$, ($CP'$ being the perpendicular from C upon $A'B'$), and similar values will hold for the sines of the other angles, such as ACB. Substituting these values in the equation already arrived at, a result will be found of the same form as the given equation, with $A'B'$ in place of AB, and the projections of the other lines also in place of the lines themselves. *Q. E. D.*

The principle just established furnishes another proof of the anharmonic properties of a pencil (see Arts. 15 and 16). This will readily appear by considering, in the figure of Article 15, $A'B'$ or $A''B''$ as the projection of AB, V being the centre of projection, and supposing $\dfrac{AD \cdot BC}{AB \cdot CD} = k$.

111. Before proceeding to give examples on the foregoing principles, there is still one important remark to be made, viz:—

Two right lines which intersect will be projected into parallel lines when the plane of projection is parallel to the right line joining the centre of projection to the point of intersection; for, the point in the projected figure, corresponding to the point of intersection, being at infinity, the projected lines are parallel.

If the given figure contain *several sets of intersecting lines, whose respective points of intersection lie in one right line,* they will become in the new figure *so many sets of parallel lines,* provided that the plane of projection be parallel to that determined by the right line before mentioned, and the centre of projection. For example, *any quadrilateral can be projected into a parallelogram* by taking the plane of projection parallel to that which passes through the centre of projection, and through the third diagonal of the quadrilateral.

From what has been said it follows *that all the points at infinity in a given plane* may be regarded as the projections of those of a single right line, and therefore may be said to be themselves *in one right line* (Poncelet, Traité des Propriétés Projectives, p. 53).

EXAMPLES ON PROJECTION.

112. The advantage of the method of projection is, that by its means certain properties of figures may be transferred immediately to other figures of a *more general* nature. Thus, for instance, certain properties of plane and spherical quadrilaterals may be said to be included in those of a parallelogram, and in like manner (see Chapter XII.), many general theorems relating to plane and spherical conics follow at once from known properties of the circle. We shall at present confine ourselves to such examples of the method as are most likely to be interesting to the class of students for which this work is more particularly intended.

1°. "If a spherical quadrilateral be divided into any two others, the *arc** joining the points of intersection of the diagonals of the partial quadrilaterals passes through the intersection of those of the original." This is evident from Articles 19 and 107, by projecting the spherical figure upon any plane.

2°. "If the sides of a spherical triangle pass respectively through three given points which lie on a great circle, and if two angles move on two given great circles, the third angle will move on one of two great circles passing through the intersection of the two former." This follows from Article 21, in the same way as before.

3°. "If a spherical hexagon be inscribed in a lesser circle, the three points of intersection of the opposite sides lie on a great circle." This follows from Articles 27 and 108, by projecting the spherical figure on a plane touching the sphere at the pole (or *spherical centre*) of the lesser circle.

4°. "If a spherical hexagon be circumscribed to a lesser circle, the three arcs joining the opposite vertices meet in a point." This follows from Art. 42, Prop. 1°, in the same way as before.

113. The properties involved in the preceding examples are of the class called by Poncelet *graphical*, as referring only to the relative positions of points and lines. We shall now give

---

* By the term "arc" we shall in general imply an arc *of a great circle.*

some which include the magnitude of lines, and belong to the class which he calls *metrical*.

1°. "If from the angles of a spherical triangle three arcs be drawn meeting in a point, and cutting the opposite sides, the products of the *sines* of the alternate segments of the sides are equal."

Take the projection of the spherical figure on any plane, and we shall have a plane triangle with three right lines drawn from its angles meeting in a point, and therefore (Art. 9, Lemma 1) the products of the alternate segments of the sides are equal. We have, then, an equation of the kind mentioned in Article 110, the number k being in the present instance 1. The Proposition is evident, therefore, from the second part of that Article. A direct proof of this Proposition will be found in Art. 125.

2°. "If a great circle cut the three sides of a spherical triangle, the products of the *sines* of the alternate segments of the sides are equal."

This follows from Art. 9, Lemma 2, by Art. 110, just as before. This Proposition also is proved directly in Art. 125.

3°. Let PQRS be a parallelogram, and OM, LN parallel to the sides. It is evident that the product $PL \cdot QM \cdot RN \cdot SO$ = the product $PO \cdot SN \cdot RM \cdot QL$. Now let us consider this parallelogram as the projection of a quadrilateral (Art. 111), and we shall have the following property:—

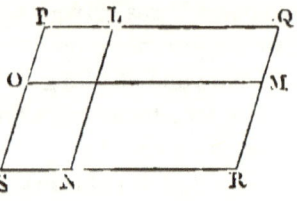

If from the extremities of the third diagonal of any plane quadrilateral two right lines be drawn, each cutting a pair of opposite sides, the products of the alternate segments so formed on the sides are equal.

And from this, again (Art. 110), *is deduced a similar Proposition for a spherical quadrilateral, the sines of the segments being taken* in the enunciation.

4°. *From a given point O on the surface of a sphere let a great circle be drawn cutting a number of given great circles in A, B, C, &c., and let a point X be taken on it, such that* $\cot OX = \cot OA + \cot OB + \cot OC + \&c.$; *required the locus of X.*

Taking the radius of the sphere to be unity (as in general we shall do), let us conceive a plane to be drawn touching the sphere at the point O, and the spherical figure of the proposed question to be projected on the tangent plane. We shall then have a number of given right lines corresponding to the given great circles, and a revolving right line corresponding to the revolving great circle (Art. 107). The arcs OX, OA, OB, &c., will be projected into their tangents, which we may express by OX', OA', OB', &c., and the condition of the question will become $\dfrac{1}{OY'} = \dfrac{1}{OA'} + \dfrac{1}{OB'}$ + &c., The locus of the point X' in the tangent plane is therefore (Art. 14) a right line, and therefore the locus of X is *a great circle*.

The examples given in this and the last Articles will serve to give the student some idea of the use of the method of projection. Many other instances of its application cannot fail to suggest themselves to the reader, who is familiar with the earlier portion of this work, and we shall have occasion still further to exemplify its principles as we proceed.

114. It is easy to see that *the projection of any plane figure on a parallel plane is similar to the original figure.* If, for example, the given figure be a circle, the projection is also a circle, whose centre corresponds to the centre of the given circle.

It is also possible " to project a circle into a circle on a plane *not parallel* to the original." This will appear from the following theorem:—

" Let AB be the diameter of the circular base of an oblique cone, whose vertex is C, and let CAB be a triangle representing the *principal section* of the cone (that is, a section whose plane passes through the vertex and the centre of the base, and is perpendicular to the plane of the base). Let PQR represent a plane section perpendicular to the plane ACB, and such that the angle PCQ = the angle CAB; then PQR will be a circle."

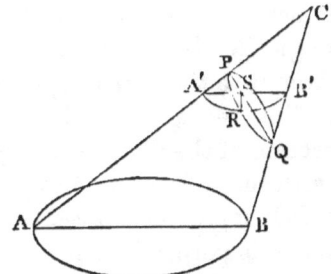

In order to prove this Proposition, let us take any point R on the section PQR, and through R draw a plane A'B'R parallel to the base of the cone. This latter plane will cut the cone in a circle, whose diameter is A'B', and will cut the plane of PQR in a right line, SR, perpendicular to the plane of ACB (because when two planes are perpendicular to a third, their intersection is also perpendicular to it). As SR is perpendicular to the plane of ACB, it is perpendicular to every line in the plane, and therefore perpendicular to PQ, and to A'B'; and as A'B'R is a semicircle, $SR^2 = A'S \cdot B'S$ (Lardner's Euclid, B. ii. Prop. 14). Now the angle $CQP = CAB = CA'B'$; therefore $PS \cdot SQ = A'S \cdot SB' = SR^2$; and as this is true for any point on the section PQR, it follows that that section is a circle, whose diameter is PQ. *Q. E. D.* (A section of an oblique cone, such as we have been considering, is called a *subcontrary* section in relation to the base.)

PROJECTION OF ANGLES.

115. The following principle is of use in questions relating to the projection of *angles* :—

Two right lines forming an angle will be projected into two others containing an equal angle, when the line joining the centre of projection to the vertex of the given angle makes equal angles in opposite directions with the plane of projection and the plane of the given angle, and when it is also perpendicular to the intersection of the two planes.

We shall first suppose the vertex of the given angle to lie in the plane of projection. Describe a sphere, with the vertex V of the given angle AVB as centre, and the line from it to the centre of projection, C, as radius (see fig.). Let PV be the intersection of the plane of the given angle with that of projection PA'B', and let COO' represent a great circle, whose plane is perpendicular to the line PV, and therefore perpendicular to the two former planes. 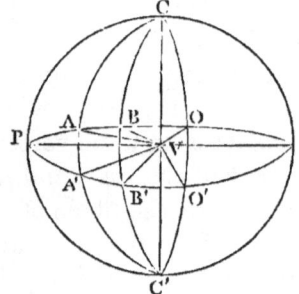 (This plane will pass through C, since by hypothesis CVP is a right angle.) Now, from the other condition granted, it is evident that the angle $CVO = C'VO'$, and therefore the arc $CO = C'O'$, and the two spherical triangles COB, C'O'B', being right-angled at O and O', and having the angles BCO, B'C'O' also equal, are equal in all respects, so that the arc $OB = O'B'$; and in the same way $OA = O'A'$, and therefore the arc $AB = A'B'$, and consequently the angle $AVB =$ the angle $A'VB'$. *Q. E. D.*

If the vertex of the given angle does not lie in the plane of projection, draw through it a plane parallel to that plane, and then the given angle will be equal to its projection on the plane so drawn (by what has been proved), and therefore equal to its projection on the original plane. (By the projection of an angle is meant the angle under the projections of the legs of the angle. It is evident that *the projection is equal to the angle itself when the plane of projection is parallel to that of the angle.*)

CHAPTER VIII.

SPHERICAL PENCILS AND SPHERICAL INVOLUTION.

116. WE shall commence this Chapter with the anharmonic properties of four planes intersecting in a right line, the connexion of which with the theory of *spherical pencils* will presently be seen.

If four given planes intersect in the same right line, and be cut by any right line in the points A, B, C, D, *the anharmonic ratio of the four points* (see Art. 15) *is constant.* (This we shall call *the anharmonic ratio of the four planes* A, B, C, D, the same letter always indicating a point on the same plane.)

In order to prove this Proposition, let us draw a second right line cutting the planes in the points A', B', C', D', and through each of the transversals AD, A'D', let a plane be drawn. The two planes so drawn will intersect in a right line, which will cut the four given planes in four points A'', B'', C'', D'', whose anharmonic ratio is the same as that of A, B, C, D (Art. 15), and also the same as that of A', B', C', D'; the two latter anharmonic ratios are therefore equal, and the Proposition immediately follows.

The anharmonic ratio of four planes is evidently the same as that of the pencil formed by any fifth plane cutting them.

If we draw a plane perpendicular to the line of intersection of the four given planes we shall have a pencil whose angles are respectively equal to those contained by the planes, and thus *we can express the anharmonic ratio of four planes by means of their mutual inclinations.* Expressing the angle between the planes A, B by the symbol (A, B), and the other angles, in a similar way, we have (Note to Art. 16) the anharmonic ratio

$$\frac{AD \cdot BC}{AB \cdot CD} = \frac{\sin(A, D) \cdot \sin(B, C)}{\sin(A, B) \cdot \sin(C, D)}.$$

117. "If the four planes be such that *one* line AD is cut harmonically by them, we have for all such lines $\frac{AD \cdot BC}{AB \cdot CD} = 1$, that is, *all* those lines are cut harmonically." (The planes A and C are here *conjugate* (see Art. 2), and also B and D.)

"When three of the planes forming an *harmonic system* are given (a definite pair being *conjugate*), the fourth is determined."

For, if any right line be drawn cutting the three given planes, we shall have on this line three given points of a line cut harmonically (a definite pair being conjugate), and therefore (Art. 3) the fourth point is determined. A plane passing through this point and the line of intersection of the given planes will be the fourth plane required.

The following problem will serve to exemplify the use of the principle just proved:—

From a given point O let a right line be drawn cutting two given planes in the points A, B, and let a distance, OX be taken on it, such that $\frac{1}{OX} = \frac{1}{OA} + \frac{1}{OB}$; *required the locus of the point* X.

Draw a plane through O and the line of intersection of the two given planes, and find its harmonic conjugate; a plane parallel to the plane so found, and bisecting a right line joining O to any point in the line of intersection of the given planes, will be the locus required. The proof exactly corresponds to that given in Art. 13; and remarks similar to those made in that Article apply in the present case.

"If *any number* of planes be given, and we have $\frac{1}{OX} = \frac{1}{OA} + \frac{1}{OB} + \frac{1}{OC} +$ &c., the locus *is still a plane*." The proof is similar to that in Article 14, to which the reader is referred.

118. *If four given great circles be drawn from the same point, V, on the surface of a sphere, and be cut by any fifth great circle in the points* A, B, C, D (see fig.) *the ratio expressed by the fraction* $\dfrac{\sin AD \cdot \sin BC}{\sin AB \cdot \sin CD}$ *is constant.*

For (Note to Art. 16), this fraction expresses the anharmonic ratio of the pencil formed by joining the points A, B, C, D to the centre of the sphere, which (Art. 116) is constant, being *the same as the anharmonic ratio of the planes of the four given great circles.*

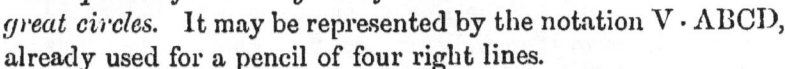

This constant ratio we shall call *the anharmonic ratio of the spherical pencil formed by the four great circles.* It may be represented by the notation V · ABCD, already used for a pencil of four right lines.

The anharmonic ratio of a spherical pencil may be expressed by means of the angles made by the great circles. For, the angle under two great circles being the same as that contained by their planes, we have (Art. 116) $\dfrac{\sin AVD \cdot \sin BVC}{\sin AVB \cdot \sin CVD}$ for the required expression.

ANHARMONIC RATIO ON THE SPHERE.

119. As it is usual to treat of spherical geometry in a manner independent of the principles of projection, we shall occasionally give direct proofs of some of the fundamental Propositions arrived at by that method. In some instances the proofs derived from the formulæ of spherical trigonometry will, perhaps, appear to the student preferable to those derived by the process of projection. The latter method, however, (wherever it is applicable), is undoubtedly the more scientific, as it is, in fact, a

method of *discovery*, which must always be carefully distinguished from a mere process of *verification*. We shall now demonstrate a general principle of spherical geometry, which furnishes a direct proof of the particular results given in the last Article.

"If in a figure on the sphere we have the product of the sines of one set of arcs divided by the product of the sines of another set $= k$, each arc of one set being part of the same great circle with one of the other set, and the whole system of extremities of arcs in both sets being the same, then the given equation will be true, when in place of the arcs, we substitute the *spherical angles* subtended by them at any point on the sphere."

Let AB be one of the arcs, and V the point assumed on the sphere; let VP be an arc perpendicular to AB;
we have then $\sin AB = \dfrac{\sin AV \cdot \sin AVB}{\sin ABV}$; but

$\sin ABV = \dfrac{\sin VP}{\sin VB}$; therefore

$$\sin AB = \frac{\sin AV \cdot \sin BV \cdot \sin AVB}{\sin VP}.$$

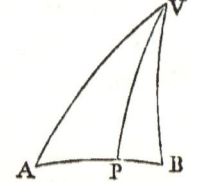

Substituting in the given equation this expression for sin AB, and similar values for the sines of the other arcs, such as AB, it will be seen from the conditions granted, that all the arcs such as VA, VB, VP, will disappear, and the result will be as stated in the enunciation above given.

Hence, in the case of a spherical pencil (see fig. in Art. 118), the equation.

$$\frac{\sin AD \cdot \sin BC}{\sin AB \cdot \sin CD} = k, \text{ becomes } \frac{\sin AVD \cdot \sin BVC}{\sin AVB \cdot \sin CVD} = k,$$

from which follows

$$\frac{\sin AD \cdot \sin BC}{\sin AB \cdot \sin CD} = \frac{\sin AVD \cdot \sin BVC}{\sin AVB \cdot \sin CVD} = \text{constant}.$$

In a similar way it appears that

$$\frac{\sin AC \cdot \sin BD}{\sin AB \cdot \sin CD} = \frac{\sin AVC \cdot \sin BVD}{\sin AVB \cdot \sin CVD} = \text{constant}.$$

120. We may here remark that *when three legs of a spherical pencil are given along with its anharmonic ratio, the fourth leg also is determined* (its relative position being supposed known as in Art. 17).

For the question is always reducible to this:—Given the sum or difference of two angles, and the ratio of their sines, to find the angles. Now, knowing the ratio of the sines, we know the ratio of their sum to their difference, which gives the ratio of tan ½ (sum of the angles) to tan ½ (their difference). Hence the angles can be found.

The position of the fourth leg may be determined *geometrically* by Art. 17, from the consideration that the anharmonic ratio of four great circles is the same as that of their four planes, which appears from Art. 118.

Since the sine of an angle is the same as the sine of its supplement it is evident that *the anharmonic ratio of a spherical pencil remains unchanged when the productions* (through V) *of any of the arcs* VA, VB, VC, VD, *are cut by the fifth arc*, the ratio being always written as in Art. 118, and the same letter, A, B, C, or D, being used to express a point anywhere on the same leg. For instance (see figure in Art. 118),

$$\frac{\sin A'D \cdot \sin B'C'}{\sin A'B \cdot \sin C'D'} = \frac{\sin A'VD' \cdot \sin B'VC'}{\sin A'VB' \cdot \sin C'VD'} = \frac{\sin AVD \cdot \sin BVC}{\sin AVB \cdot \sin CVD}$$
$$= \frac{\sin AD \cdot \sin BC}{\sin AB \cdot \sin CD}.$$

What has been said agrees with the rule already laid down in Art. 15, to which the reader is referred, and the other remarks there made admit of like application on the sphere.

By the expression, *anharmonic ratio on four points*, A, B, C, D, *on a great circle*, we are to understand that of a spherical pencil, whose vertex is anywhere on the surface of the sphere, and whose legs are four great circles passing through them. This is not to be confounded with the anharmonic ratio of four points on a circle considered as lying *in plano*, such as we had in Art. 25. In what follows we shall use the expression in the sense above explained.

HARMONIC PROPERTIES ON THE SPHERE.

121. When the planes of the four great circles form an harmonic system (see Art. 117), the pencil is called *a spherical harmonic pencil*. In this case the radii drawn to the four points, A, B, C, D

(see figure in Art 118) make a plane harmonic pencil, and the arc AD is said to be cut harmonically. This gives, for an arc so cut (Note to Art. 16), $\sin AD \cdot \sin BC = \sin AB \cdot \sin CD$. It follows from Art. 118 that *if one arc, AD, be cut harmonically by a spherical pencil* $\sin AVD \cdot \sin BVC = \sin AVB \cdot \sin CVD$, *and therefore all transverse arcs, such as AD, will be cut harmonically, and the pencil itself will be harmonic.*

It is hardly necessary to observe that the transversal may intersect the productions of any of the arcs through V; or that when three of the legs of a spherical harmonic pencil are given, a definite pair being *conjugate* (see Art. 117), the fourth is determined.

122. *When an arc, AD, is cut harmonically* (see Art. 121), *the tangents of* AB, AC, AD, *are in harmonic proportion.*

For, let the right line AD' (see fig.) be the intersection of the plane of the great circle AD with the plane touching the sphere at A; then, as the radius of the sphere, AO, is unity, AB', AC', AD', are the tangents of AB, AC, AD; now AD' is cut harmonically (since the four radii OA, OB, OC, OD, form an harmonic pencil), and therefore AB', AC' AD', are in harmonic proportion.

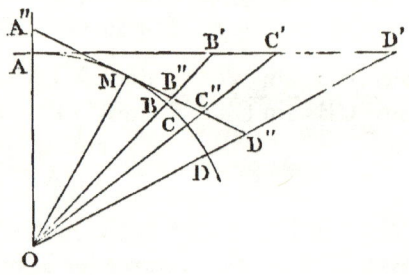

We shall now prove the same result directly from the trigonometrical relation:—

Since (Art. 121)
$$\sin AD \cdot \sin BC = \sin AB \cdot \sin CD,$$
we have
$$\sin AD \cdot \sin (AC - AB) = \sin AB \cdot \sin (AD - AC);$$
therefore
$$\sin AD \cdot \sin AC \cdot \cos AB - \sin AD \cdot \cos AC \cdot \sin AB =$$
$$\sin AB \cdot \sin AD \cdot \cos AC - \sin AB \cdot \cos AD \cdot \sin AC,$$
and, dividing both sides by
$\sin AB \cdot \sin AC \cdot \sin AD$, $\cot AB - \cot AC = \cot AC - \cot AD$, that is, the cotangents of AB, AC, AD, are in arithmetic propor-

tion, and therefore (Art. 5, 3°) the tangents are in harmonic proportion. Q. E. D.

Again, by drawing a tangent plane at C, we shall have (Art. 5, 3°)
$$2 \cot CA = \cot CB - \cot CD.$$

This also follows readily from the trigonometrical relation:—
For,
$$\sin(AC + CD) \cdot \sin BC = \sin(AC - BC) \cdot \sin CD,$$
that is,
$$(\sin AC \cdot \cos CD + \cos AC \cdot \sin CD) \sin BC =$$
$$(\sin AC \cdot \cos BC - \cos AC \cdot \sin BC) \sin CD,$$
and, dividing both sides by
$\sin AC \cdot \sin BC \cdot \sin CD$, $\cot CD + \cot AC = \cot BC - \cot AC$,
therefore,
$$2 \cot AC = \cot BC - \cot DC, \text{ as before.}$$

123. If a tangent plane be drawn at the middle point, M, of the mean arc, AC (see preceding figure), the right line $A''D''$, in which it intersects the plane of the great circle AD, will be cut harmonically by the four radii mentioned in the last Article, and, as MA'' evidently $= MC''$, we shall have (Art. 3) MB'', MC'', MD'', in geometric proportion. Hence, *when an arc is cut harmonically, and either of the mean arcs is bisected, the tangents of the three arcs, measured from the point of bisection to the other points of section, are in geometric proportion.*

Conversely, it is evident that if $\tan MB \cdot \tan MD = \tan^2 MC$, and MA be taken equal to MC, the whole arc AD will be cut harmonically.

The preceding results may be immediately deduced from the trigonometrical condition of harmonic section. To prove the first, for instance, we have
$$\sin AD \cdot \sin BC = \sin AB \cdot \sin CD,$$
or,
$$\frac{\sin AD}{\sin CD} = \frac{\sin AB}{\sin BC};$$
therefore,
$$\frac{\sin AD - \sin CD}{\sin AD + \sin CD} = \frac{\sin AB - \sin BC}{\sin AB + \sin BC},$$

from which,
$$\frac{\tan\tfrac{1}{2}(AD-CD)}{\tan\tfrac{1}{2}(AD+CD)} = \frac{\tan\tfrac{1}{2}(AB-BC)}{\tan\tfrac{1}{2}(AB+BC)},$$
and therefore
$$\frac{\tan MC}{\tan MD} = \frac{\tan MB}{\tan MC},$$
or,
$$\tan^2 MC = \tan MB \cdot \tan MD.$$

124. *If the angle made by two great circles be bisected internally and externally, the two great circles and the two bisecting arcs form a spherical harmonic pencil.*

For, let VA and VC be the given arcs (see fig. of Art. 118), and VB, VD, the bisecting arcs; then we have sin AVD = sin A'VD = sin CVD, and sin AVB = sin BVC, and therefore sin AVD · sin BVC = sin AVB · sin CVD. Hence (Art. 121) the pencil is harmonic.

Conversely, *if in a spherical harmonic pencil two alternate legs be at right angles, they are the bisectors of the angles made by the other pair.*

For, as
$$\sin AVD \cdot \sin BVC = \sin AVB \cdot \sin CVD,$$
we have
$$\frac{\sin AVD}{\sin AVB} = \frac{\sin CVD}{\sin BVC},$$
or (since BVD is a right angle),
$$\frac{\sin A'VD}{\cos A'VD} = \frac{\sin CVD}{\cos CVD};$$
that is, tan A'VD = tan CVD, and therefore A'VD = CVD, whence AVB = CVB.

125. The harmonic properties of a spherical triangle follow at once by projection from those of a plane triangle given in Art. 9. For the reason stated in Art. 119, we shall give here direct demonstrations of the properties in question.

LEMMA 10. If three arcs of great circles be drawn from the angles of a spherical triangle to meet in one point, the products of the sines of the alternate segments so made on the sines are equal.

HARMONIC PROPERTIES ON THE SPHERE.

Let ABC be the spherical triangle, and O the point (see fig.); then,

$$\frac{\sin AB'}{\sin AO} = \frac{\sin AOB'}{\sin AB'O}$$

$$\frac{\sin CA'}{\sin CO} = \frac{\sin COA'}{\sin CA'O}$$

and

$$\frac{\sin BC'}{\sin BO} = \frac{\sin BOC'}{\sin BC'O};$$

therefore, by multiplication,

$$\frac{\sin AB' \cdot \sin CA' \cdot \sin BC'}{\sin AO \cdot \sin CO \cdot \sin BO} = \frac{\sin AOB' \cdot \sin COA' \cdot \sin BOC'}{\sin AB'O \cdot \sin CA'O \cdot \sin CB'O};$$

and, in a similar manner,

$$\frac{\sin A'B \cdot \sin C'A \cdot \sin B'C}{\sin BO \cdot \sin AO \cdot \sin CO} = \frac{\sin A'OB \cdot \sin C'OA \cdot \sin B'OC}{\sin BA'O \cdot \sin AC'O \cdot \sin CB'O};$$

the right-hand members of these two equations are evidently equal, therefore so are those on the left hand, and therefore we have $\sin AB' \cdot \sin CA' \cdot \sin BC' = \sin A'B \cdot \sin C'A \cdot \sin B'C$.

The converse of this Lemma may be employed (under a restriction similar to that in Art. 9, Lemma 1) to prove that arcs drawn from the angles of a spherical triangle meet in one point.

LEMMA 11. If a great circle be drawn cutting the three sides of a spherical triangle, the products of the sines of the alternate segments so formed are equal.

Let ABC be the triangle (see fig.), and B'A'C' the transversal.

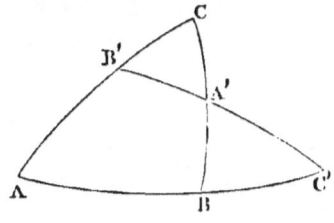 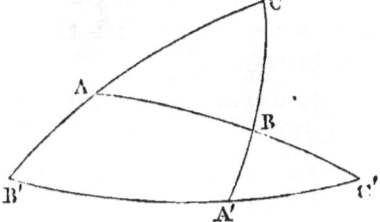

We have

$$\frac{\sin AB'}{\sin AC'} = \frac{\sin AC'B'}{\sin AB'C'}, \quad \frac{\sin CA'}{\sin CB'} = \frac{\sin CB'A'}{\sin CA'B'},$$

and
$$\frac{\sin BC'}{\sin BA'} = \frac{\sin BA'C'}{\sin BC'A'};$$

therefore, by multiplication,

$$\frac{\sin AB' \cdot \sin CA' \cdot \sin BC'}{\sin AC' \cdot \sin CB' \cdot \sin BA'} = \frac{\sin AC'B' \cdot \sin CB'A' \cdot \sin BA'C'}{\sin AB'C' \cdot \sin CA'B' \cdot \sin BC'A'}.$$

Now, the right-hand member of this equation evidently $= 1$, therefore, the left-hand member also $= 1$, and therefore

$\sin AB' \cdot \sin CA' \cdot \sin BC' = \sin A'B \cdot \sin C'A \cdot \sin B'C.$ *Q. E. D.*

Since (by the Proposition just proved) $\dfrac{\sin AC'}{\sin BC'} = \dfrac{\sin AB' \cdot \sin CA'}{\sin A'B \cdot \sin B'C}$

it follows that an arc joining the middle points of the sides of a spherical triangle cuts the base externally in such a manner that the segments, AC', BC' (see fig.), are supplemental.

The converse of Lemma 11 is of use on the sphere in a way analogous to that in which the converse of Lemma 2 of Art. 9 is employed *in plano*.

126. If now (see figure in Art. 125, Lemma 10) we join A'B', B'C', C'A', by arcs of great circles, *all the arcs on the figure will be cut harmonically, and the points* A'', B'', C'' *will lie on a great circle.* (There are, of course, *two* points A'', and two B'', and two C''.)

For, by Lemma 11, we have

$$\frac{\sin AC''}{\sin BC''} = \frac{\sin AB' \cdot \sin CA'}{\sin A'B \cdot \sin B'C} = \frac{\sin C'A}{\sin BC'} \text{ (Lemma 10);}$$

therefore, $\sin AC' \cdot \sin BC'' = \sin C''A \cdot \sin BC'$, that is (Art. 121), the arc AC'' is cut harmonically. And in a similar way it may be proved, that CA'' and AB'' (or CB'', see figure) are cut harmonically.

Again, as AC'' is cut harmonically, by joining CC'', we see that B'C'' is (Art. 121) cut harmonically, and for a similar reason B'A'' and C'B'' (or A'B'') are cut harmonically. Also, as CA'' is cut harmonically, so is CC'; for a similar reason AA', and BB'.

Lastly, taking CC'', AC'', B'C'' as legs of an harmonic pencil (the first pair being conjugate), the fourth leg must (Art. 121) pass through A'' and B'', that is, A'', B'', C'' lie on the same great circle.

127. From the preceding Article, we may deduce the following remarkable property:—

If two of the three diagonals of a spherical quadrilateral be quadrants, the third also is a quadrant.

For (see figure in Art. 125, Lemma 10), let $CB'OA'$ be the quadrilateral, and let CO and $B'A'$ be quadrants, and let P be their intersection. Then, since $B'C''$ is cut harmonically, we have (Art. 122) $\cot B'C'' + \cot B'P = 2 \cot B'A' = 0$; therefore $B'C''$ is the supplement of $B'P$, and therefore their sines are equal, and consequently $\sin A'C'' = \sin A'P$, whence finally, $A'C'' = A'P$. In like manner we find $CO' = OP$. Now let us take AOA' as a great circle cutting the sides of the triangle $C'PC''$; then (Lemma 11 of Art. 125) AC'' is the supplement of AC', and, as AC'' is cut harmonically, $2 \cot AB = \cot AC' + \cot AC'' = 0$; therefore $AB =$ a quadrant.

In a similar manner we can prove the proposition for either of the other diagonals.

We shall conclude what we have to say on the harmonic properties of a spherical triangle by remarking, that results strictly analogous to those contained in Articles 10, 11, and 12, hold good on the sphere. These results may be proved either by projection, or directly by the principles of spherical geometry, above given.

ANHARMONIC PROPERTIES OF A LESSER CIRCLE.

128. We have, in the next place, to explain the fundamental anharmonic properties of lesser circles on the sphere. We shall deduce these, in the first instance, from the *anharmonic property of a cone* (see Art. 108).

If through four fixed sides of a cone four planes be drawn intersecting in any fifth variable side, the anharmonic ratio of the four planes is constant.

For, by Art. 116, the anharmonic ratio of the four planes is the same as that of the pencil formed by their intersections with the circular base of the cone, and the latter anharmonic ratio is constant, by Art. 25.

Suppose now four arcs to be drawn from four fixed points on a lesser circle to any variable fifth point on the circle, it fol-

lows, from what has been just proved, that *the anharmonic ratio of the spherical pencil so formed is constant*, since this ratio is the same (Art. 118) as that of the planes of the four arcs. (This constant ratio we shall call *the anharmonic ratio of the four points on the lesser circle.*)

We can express this constant ratio in terms of the sides of the spherical quadrilateral, of which the four fixed points are corners.

Let A, B, C, D be the fixed points, and V the fifth point. It follows from Art. 26, that the ratio in question is expressed by the product of the chords of AD and BC, divided by the product of the chords of AB and CD. But the chord of an arc is equal to twice the sine of half the arc. Therefore the anharmonic ratio of the pencil

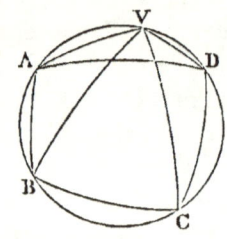

$$= \frac{\sin\tfrac{1}{2} AD \cdot \sin\tfrac{1}{2} BC}{\sin\tfrac{1}{2} AB \cdot \sin\tfrac{1}{2} CD}.$$

129. The last result may be proved directly in the following manner:—

If a, b, c be the sides of a spherical triangle, R the spherical radius of the lesser circle circumscribed to the triangle, and C the angle opposite to the side c, we have the known formulæ,

$$\sin C = \frac{2N}{\sin a \cdot \sin b} \text{ and } \tan R = \frac{2 \sin\tfrac{1}{2} a \cdot \sin\tfrac{1}{2} b \cdot \sin\tfrac{1}{2} c}{N},$$

N expressing
$$\sqrt{\{\sin s \cdot \sin(s-a) \cdot \sin(s-b) \cdot \sin(s-c)\}},$$
in which
$$s = \tfrac{1}{2}(a+b+c).$$

By multiplying these values of sin C and tan R, we find

$$\tan R \cdot \sin C = \frac{4 \sin\tfrac{1}{2} a \cdot \sin\tfrac{1}{2} b \cdot \sin\tfrac{1}{2} c}{\sin a \cdot \sin b} = \frac{\sin\tfrac{1}{2} c}{\cos\tfrac{1}{2} a \cdot \cos\tfrac{1}{2} b},$$

and therefore
$$\sin C = \frac{\cot R \cdot \sin\tfrac{1}{2} c}{\cos\tfrac{1}{2} a \cdot \cos\tfrac{1}{2} b}.$$

Now, applying this result to the triangles AVD, BVC (see last figure), we get, by multiplication,

$$\sin AVD \cdot \sin BVC = \frac{\cot^2 R \cdot \sin\tfrac{1}{2} AD \cdot \sin\tfrac{1}{2} BC}{\cos\tfrac{1}{2} AV \cdot \cos\tfrac{1}{2} DV \cdot \cos\tfrac{1}{2} BV \cdot \cos\tfrac{1}{2} CV}.$$

In a similar manner, from the triangles AVB, CVD,

$$\sin AVB \cdot \sin CVD = \frac{\cot^2 R \cdot \sin\tfrac{1}{2} AB \cdot \sin\tfrac{1}{2} CD}{\cos\tfrac{1}{2} AV \cdot \cos\tfrac{1}{2} BV \cdot \cos\tfrac{1}{2} CV \cdot \cos\tfrac{1}{2} DV};$$

and, by division,

$$\frac{\sin AVD \cdot \sin BVC}{\sin AVB \cdot \sin CVD} = \frac{\sin\tfrac{1}{2} AD \cdot \sin\tfrac{1}{2} BC}{\sin\tfrac{1}{2} AB \cdot \sin\tfrac{1}{2} CD}. \quad Q.E.D.$$

SPHERICAL INVOLUTION.

130. The attentive reader will at once perceive that the anharmonic properties of spherical pencils given in this Chapter admit of applications analogous to those already explained at some length with respect to pencils *in plano*. The theorems and problems alluded to require in general only modifications of a very obvious nature, in order to become adapted to the sphere. We shall now, therefore, pass to the principles of spherical involution.

If three pairs of points AA', BB', CC', on a great circle, be such that the anharmonic ratio of four of the points (see Art. 120) *is the same as that of their four conjugates (the relative order of the two sets being also the same), then, every set of four will possess a similar property.* (Three such pairs of points are said to be in *spherical involution*.)

Taking the projection of the great circle on any plane, we have (Arts. 110 and 35) three pairs of points *in a right line* forming an involution. The Proposition follows from this by Art. 110.

It is evident from what has been said, that the two properties of rectilinear involution, given in Art. 76, may be transferred to the sphere by substituting the *sines* of the spherical distances in place of the segments of the right line.

By this mode of proof we can also immediately ascertain the existence of an *infinite system of points* in spherical involution. For it is evident that two of the three pairs above mentioned being given, the third pair may vary indefinitely, (because (Art. 36) their projections on the right line may do so), and that any three pairs of the entire number so formed will be in involution.

s

We see, moreover, that a point may coincide with its conjugate (because its projection may); such a point we shall call a focus. The number of these is *four*, each point in the projection corresponding to two points on the sphere; and it follows from Arts. 36 and 121, that any two of the foci not diametrically opposite, together with a pair of conjugates, form a system of harmonic points on the sphere.

131. The point bisecting the arc between two foci (selected as before) is analogous to the centre of the system *in plano*, and may be called by the same name. There are *four centres*, and any of them may be taken in the enunciation of the properties analogous to those relating to the centre in the case of a right line.

To find the properties of these points, we need only take the plane of projection (which has hitherto been arbitrary), so as to touch the sphere at one of them. In that case the point of contact will be the centre of the projected system of points (because it bisects the distance between the foci), and (calling O the centre) we find (Arts. 35 and 36) $\tan OA \cdot \tan OA'$

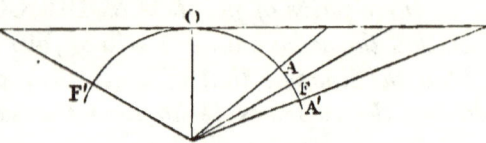

$= \tan OB \cdot \tan OB' =$ &c., that is, *the product of the tangents of the arcs between the centre and any pair of conjugates is constant.*

When the foci (F, F') are real (see fig.), we shall have of course $\tan OA \cdot \tan OA' = \tan^2 OF$. This result is deducible from Art. 123 also.

As $\tan OA \cdot \tan OA'$ is constant, AA' being any pair of conjugates, it follows that if OA vanishes, $OA' = 90°$. Hence, *a centre of a system of points in spherical involution is the conjugate of a point at the distance of a quadrant.**

---

* It is to be observed that OA and OA' are to be measured in the same direction from O when the foci are real, and in opposite directions when they are imaginary.

With respect to spherical involution, Mr. Davies appears to have fallen into a mistake (see "Mathematician," vol. i. p. 248). From what is there said it would follow that we should have $\tan^2 \frac{1}{2} OF = \tan\frac{1}{2} OA \cdot \tan\frac{1}{2} OA' = \tan\frac{1}{2} OB \cdot \tan\frac{1}{2} OB' =$ &c. instead of the equations given in the text.

132. Three pairs of arcs drawn from a point on the sphere are said to form *a pencil in spherical involution* when any transversal arc is cut by them in involution. The following are examples:—

1°. "Six arcs drawn from any point on the surface of the sphere to the corners of a spherical quadrilateral, and to the extremities of its third diagonal, make a pencil in involution."

Project the figure on any plane. We have then (Art. 37, 1°) a plane pencil in involution. It follows from this (Art. 116) that the *planes* of the great circles forming the spherical pencil may be said to be *in involution*, and therefore (Art. 118) the arcs themselves are so.

The Proposition may of course be proved directly from the principles of spherical geometry by adapting the demonstration, in Art. 37, 1°, to the sphere. This will present no difficulty to the student.

2°. "If three great circles pass through the same point, and cut a given lesser circle, the six arcs drawn from the six points of section to any point on the lesser circle form a pencil in involution."

This may be proved from Art. 37, 4°, by projecting the spherical figure on the plane of the lesser circle, or directly by adapting the proof there given to the sphere by means of the anharmonic properties established in Arts. 120 and 128.

133. We shall conclude this Chapter with an example of an infinite system of points in spherical involution. We must premise the following Lemma:—

LEMMA 12.—*Through a given point O on the surface of the sphere let any arc*, OA, *be drawn cutting a given lesser circle in the points* A, A′, *the product of the tangents of the halves of the segments* OA, OA′ *is constant.*

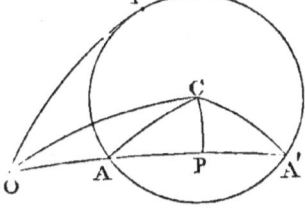

For, if the perpendicular CP (see fig.) be drawn from the spherical centre C, we have

$$\cos OC = \cos OP \cdot \cos CP,$$

and

$$\cos AC = \cos AP \cdot \cos CP;$$

therefore $\dfrac{\cos OC}{\cos AC} = \dfrac{\cos OP}{\cos AP}.$

Hence,

$\dfrac{\cos AC - \cos OC}{\cos AC + \cos OC} = \dfrac{\cos AP - \cos OP}{\cos AP + \cos OP} = \tan\tfrac{1}{2}(OP+AP) \cdot$
$\tan\tfrac{1}{2}(OP-AP) = \tan\tfrac{1}{2} OA \cdot \tan\tfrac{1}{2} OA'$ (since $AP = A'P$).

As AC and OC are given, the Proposition is proved when the point O is without the circle, and a similar proof applies when the point is within.

If OT (see last figure) be a tangent arc, we have

$$\tan\tfrac{1}{2} OA \cdot \tan\tfrac{1}{2} OA' = \tan^2 \tfrac{1}{2} OT.$$

If any number of lesser circles pass through the same two points P, Q, *and any transverse arc cut them in an indefinite number of pairs of points,* AA' BB', &c., *and also cut the arc* PQ *in the point* O, *the points of bisection of the arcs* OA, OA', OB, OB', &c., *are in spherical involution.*

For, by the Lemma,

$\tan\tfrac{1}{2} OA \cdot \tan\tfrac{1}{2} OA' = \tan\tfrac{1}{2} OP \cdot \tan\tfrac{1}{2} OQ = \tan\tfrac{1}{2} OB \cdot \tan\tfrac{1}{2} OB'$ = &c.;

therefore the Proposition is evident from Art. 131, and O is one of the centres of the system.

It is also evident that if T and T' are the points of contact of circles described through P, Q, to touch the transverse arc, the middle points of the arcs OT, OT', will be the foci of the system.

The foci are imaginary when the transversal cuts the arc PQ between P and Q. (PQ is supposed to be less than 180°.)

CHAPTER IX.

POLAR PROPERTIES OF CIRCLES ON THE SPHERE.

134. Given a point O on the sphere, and a *lesser* circle whose spherical centre is C; let CO be joined by an arc of a great circle, and on the joining arc a portion CO′ be taken (on the same side of C as O is) such that tan CO · tan CO′ = the square of the tangent of the spherical radius of the given circle, a great circle perpendicular to CO′ at the point O′ is called the *polar* of the point O, and the point O the *pole* of the perpendicular arc, *in respect to the given circle*. It follows from these definitions that if we project the figure on a plane touching the sphere at the point C, the point O 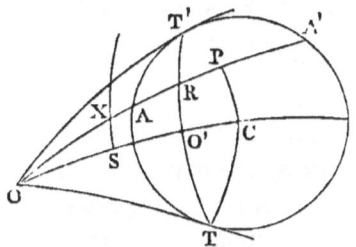 and its polar will be projected into a pole and its polar in relation to the projection of the circle on the tangent plane. This appears from Arts. 109 and 38.

From this again it follows that the same will be true if we take the projection on the plane of the lesser circle itself, or on any plane parallel to it, because projections on parallel planes are similar.

135. When the point O lies on the circumference of the lesser circle, its polar is the tangent arc drawn at the point.

When the point O is without the circle, its polar coincides with the arc joining the points of contact of great circles drawn from it to touch the lesser circle.

This is evident by projection, from what has been said in the last Article. It may be proved directly as follows:—

Let us suppose OT and OT′ to be the tangent arcs (see the preceding figure). The arc TT′ is perpendicular to CO, and cuts it in a point, which we may call O′; then, since the angle CTO is right, we have (by Napier's rules)

$$\cos TCO = \frac{\tan CO'}{\tan CT},$$

and similarly

$$\cos TCO = \frac{\tan CT}{\tan CO}; \text{ therefore, } \frac{\tan CO'}{\tan CT} = \frac{\tan CT}{\tan CO},$$

and $\tan CO \cdot \tan CO' = \tan^2 CT$, which proves the Proposition.

The reader will at once perceive that the present sense in which the word "pole" is used, differs from its common trigonometrical meaning. We shall have to use it in both senses, but a little attention to the context will be sufficient to prevent any confusion arising from this ambiguity. What is commonly called the pole of a lesser circle, we shall denominate (as we have generally done) its spherical centre.

136. *Any arc through the point* O (whether that point be within or without the circle) *is cut harmonically by its polar and by the given circle.*

For, if we project the figure on a plane touching the sphere at the spherical centre C, we shall have (Art. 39) a right line (the projection of the arc in question) cut harmonically, and therefore (Art. 121) the arc itself is cut harmonically.

The following problem will furnish a direct proof by trigonometrical formulæ:—

Given a point O *and a lesser circle on the sphere, let any arc* OA' *be drawn cutting the circle* (see fig. in Art. 134) *in* A *and* A', *and let a point* R, *the harmonic conjugate of* O, *be taken on it. Required the locus of* R.

From the spherical centre C draw an arc CP, perpendicular to OA'. This perpendicular will evidently bisect AA', and therefore (supposing O to be without the circle) $OP = \frac{1}{2}(OA' + OA)$. We have then, $\cos COP \cdot \tan OC = \tan OP = \tan \frac{1}{2}(OA' + OA) = \frac{\tan \frac{1}{2} OA' + \tan \frac{1}{2} OA}{1 - \tan \frac{1}{2} OA' \cdot \tan \frac{1}{2} OA}$ (1).

Now, from the condition of the question (Art. 122), $\cot OR = \frac{1}{2}(\cot OA + \cot OA')$; therefore, by a well-known formula, $\cot OR = \frac{1}{4}(\cot \frac{1}{2} OA - \tan \frac{1}{2} OA + \cot \frac{1}{2} OA' - \tan \frac{1}{2} OA') = \frac{1}{4}(\tan \frac{1}{2} OA' + \tan \frac{1}{2} OA)\left(\frac{1}{\tan \frac{1}{2} OA \cdot \tan \frac{1}{2} OA'} - 1\right)$ (2).

The quantity $\tan\frac{1}{2} OA \cdot \tan\frac{1}{2} OA'$ is constant (Lemma 12 of Art. 133); let it be called k, and we get (by division) from equations (1) and (2),
$$\frac{\cos COP}{\cot OR} \cdot \tan OC = \frac{4k}{(1-k)^2};$$
and therefore
$$\cos COR \cdot \tan OR = \frac{4k}{(1-k)^2} \cdot \cot OC.$$

The quantity on the right-hand side of the equation being constant, we may suppose it equal to the tangent of a known arc OO', so that we have $\cos COR \cdot \tan OR = \tan OO'$; from which it appears that an arc perpendicular to OC at the distance OO' from O (see fig.) will pass through R, and therefore this perpendicular arc is the locus required.

Since O' is a point in the locus, it is the harmonic conjugate of O on the arc through C, and therefore (Art. 123) $\tan OC \cdot \tan O'C =$ the square of the tangent of the spherical radius. *The locus in question is therefore the polar of the given point with respect to the given circle.*

If the point O be within the circle we shall have $OP = \frac{1}{2}(OA' - OA)$, and (Art. 122) $\cot OR = \frac{1}{2}(\cot OA - \cot OA')$, and the investigation will proceed just as before. The locus of R will still be the polar of O, and the Proposition stated at the commencement of this Article is therefore proved.

137. The theorem given in the last Article leads to results analogous to those given in Art. 40.

1°. "If any number of points on a great circle be taken as poles, their polars with respect to a given lesser circle pass through the same point, namely, the pole of the great circle with respect to the lesser."

Hence, *the arc joining two points has for its pole the intersection of their polars.*

2°. "If any number of great circles pass through one point, the locus of their poles with respect to a given lesser circle is a great circle, namely, the polar of the given point."

3°. "If through a given point on the sphere any two arcs be drawn cutting a given lesser circle, the arcs joining the points of intersection taken in pairs (see Art. 40, 3°), intersect on the polar of the given point."

4°. "If four arcs be drawn from the same point on the sphere cutting a lesser circle, the anharmonic ratio of any four of the points of intersection is the same as that of the remaining four, taken in the corresponding order."

5°. "If four points be taken on a great circle, their polars with respect to a lesser circle will form a spherical pencil, whose anharmonic ratio is the same as that of the four points."

These Propositions are evident by projecting the spherical figure on the plane of the lesser circle. The first four may also be proved directly from the principles already laid down. The direct proof of the fifth requires the following principle:—

LEMMA. 13.—*If four arcs be drawn from a point perpendicular to the legs of a spherical pencil respectively, the anharmonic ratio of the pencil thus formed is the same as that of the given pencil.*

Let O be the given point, and A, B, C, D the *trigonometrical* poles of the four great circles forming the given pencil, whose vertex is V. Then, since the angle between two great circles is measured by the arc joining their trigonometrical poles, the anharmonic ratio of the given pencil equals (Art. 118) that of the four poles (see Art. 120) A, B, C, D (which all lie in a great circle, a common secondary to the legs of the given pencil); but this latter ratio is the same as that of the four perpendicular arcs, since these arcs pass through the poles respectively; therefore the anharmonic ratio of the given pencil is the same as that of the four perpendicular arcs. *Q. E. D.*

By the aid of this Lemma, the adaptation of the proof given in Art. 40, 5°, to the sphere is obvious.

138. The following properties of spherical quadrilaterals are evident by projection from those given in Art. 41. They may also be deduced directly from principles already laid down.

1°. If a spherical quadrilateral be inscribed in a lesser circle, the intersection of the diagonals and the extremities of the third diagonal are three points, such that the arc joining any two is the polar of the third.

2°. If a spherical quadrilateral be circumscribed to a lesser circle, and an inscribed quadrilateral be formed by joining the successive points of contact, the diagonals of the two quadrilaterals intersect in the same point, and form a spherical harmonic pencil; and the third diagonals of the two quadrilaterals are coincident.

3°. If a spherical quadrilateral be circumscribed to a lesser circle, each of the three diagonals is the polar of the point of intersection of the other two.

4°. If a spherical quadrilateral be inscribed in, or circumscribed to a lesser circle, the arc joining the spherical centre C with the intersection of the diagonals (which we shall call V) is perpendicular to the third diagonal, and cuts in a point V′, such that $\tan CV \cdot \tan CV' =$ the square of the tangent of the spherical radius of the lesser circle.

139. We shall now give a few examples illustrating the use of polar properties on the sphere in the solution of problems.

1°. "Given a point O and a lesser circle, let any arc OA′ (see figure in Art. 134) be drawn cutting the circle, and let a portion OX be taken on it, such that $\cot OX = \cot OA + \cot OA'$; required the locus of X."

Construct the polar of O, and let R be the point where it cuts OA′. We have then $\cot OX = \cot OA + \cot OA' =$ (Arts. 136 and 122) $2 \cot OR$. Now if a perpendicular arc, XS, be drawn from X on OC, we shall have

$$\cos XOS = \frac{\tan OS}{\tan OX}, \text{ and also } \cos XOS = \frac{\tan OO'}{\tan OR};$$

therefore, $\quad \dfrac{\tan OS}{\tan OO'} = \dfrac{\tan OX}{\tan OR} = \dfrac{\cot OR}{\cot OX} = \dfrac{1}{2};$

the locus is therefore a great circle perpendicular to OC at the point S, which is determined by this last equation.

If there be several lesser circles, and OA′ cut them in the points A, A′, B, B′, C, C′, &c., and OX be taken so that $\cot OX = \cot OA + \cot OA' + \cot OB + \cot OB' + $ &c., the locus of X will still be a *great circle*, by Art. 113, 4°, to which the question is evidently reducible.

2°. "Given, as before, a point O, and a lesser circle whose

spherical centre is C, it is required to find another point O' on the sphere, such that the sines of the segments of any arc drawn through either shall be to one another as the sines of the distances from the other to the corresponding intersections of the arc with the lesser circle."

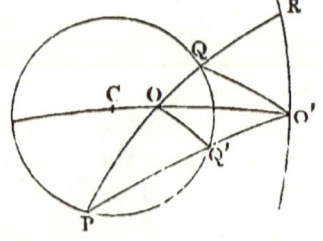

Construct the polar of the point O: then, the point O', where the arc CO (see fig.) cuts the polar, is the point required.

In order to prove this, we shall require the following Lemma:—

LEMMA 14.—*If an angle of a spherical triangle be bisected internally or externally, the sines of the segments of the opposite side are to one another as the sines of the containing sides.*

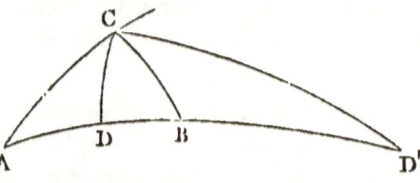

Let ABC be the triangle, and DC the internal bisector of the angle C. We have then,

$$\frac{\sin AD}{\sin AC} = \frac{\sin ACD}{\sin ADC} = \frac{\sin BCD}{\sin BDC} = \frac{\sin BD}{\sin BC};$$

therefore, $\dfrac{\sin AD}{\sin BD} = \dfrac{\sin AC}{\sin BC}.$

Again, if CD' be the external bisector,

$\dfrac{\sin AD'}{\sin AC} = \dfrac{\sin ACD'}{\sin AD'C} = \dfrac{\sin BCD'}{\sin BD'C}$ (since ACD' and BCD' are supplements) $= \dfrac{\sin BD'}{\sin BC}$; therefore, $\dfrac{\sin AD'}{\sin BD'} = \dfrac{\sin AC}{\sin BC}.$ Q. E. D.

Returning now to the former question, let PR be any arc through O; then, as PR is (Art. 136) cut harmonically, and as OO'R is a right angle, the angle PO'Q is (Art. 124) bisected, and therefore, by the Lemma, $\dfrac{\sin OP}{\sin OQ} = \dfrac{\sin O'P}{\sin O'Q}.$

Again, since the angle QO'O = Q'O'O, it is evident from the symmetry of the figure that the angle QOQ' is bisected, and therefore $\dfrac{\sin PO'}{\sin Q'O} = \dfrac{\sin PO}{\sin Q'O}.$

A similar argument applies when O is without the circle.

3°. "Given in position one pair of opposite sides of a spherical quadrilateral inscribed in a lesser circle; given also the point of intersection of the diagonals; find the locus of the spherical centre."

This question is completely analogous to that in Art. 48, 4°, and can present no difficulty to the student.

4°. *Given a lesser circle and a great circle not cutting it; required a point such that any arc being drawn through it cutting the lesser circle, the sum of the cosecants of arcs drawn perpendicular to the great circle from the points of intersection shall be constant.*

The pole of the great circle with respect to the lesser circle answers the question. In order to prove this, we shall premise the following Lemma:—

LEMMA 15.—*Given a circle* in plano *and a right line not cutting it; the sum of the reciprocals of perpendiculars on the right line from the extremities of any chord drawn through its pole is constant.*

For, let O be the pole of the given line, PQ any chord through it, and QS, PT the perpendiculars. Then, since PR is cut harmonically, RQ, RO, RP, are in harmonic proportion; therefore, QS, OO', PT, which are proportional to them, are also in harmonic proportion, and (Art. 5, 3°) $\frac{1}{QS} + \frac{1}{PT} = 2 \cdot \frac{1}{OO'}$ = constant.

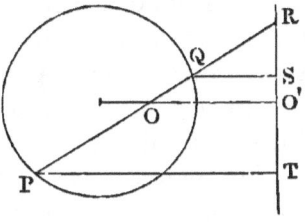

Q. E. D.

Returning now to the original question, let us project, on the plane of the lesser circle, the great circle and its pole with respect to the lesser circle, and also the arc drawn through the pole. We shall then have (Art. 134) a pole and its polar in the plane of the lesser circle, and a chord drawn through the pole, and therefore by the Lemma the sum of the reciprocals of the perpendiculars the polar from the extremities of the chord is constant; these perpendiculars multiplied by the sine of the angle between the planes of the given circles are the sines of the perpendicular arcs entering into the proposed problem; therefore, as the angle between the planes is constant, the sum of the reciprocals of the sines of the perpendicular arcs is constant, that is, the sum of the cosecants is constant.

If the given great circle intersect the lesser circle, it is the difference of the cosecants of perpendicular arcs drawn on it from the points of intersection of any arc through the pole, which will be constant.

To find the value of the constant, let A and B be the points where the perpendicular from the pole on the polar cuts the lesser circle, and O' the point where it cuts the polar; then taking the case where the arc through the pole coincides with AB, we find the constant $=$ cosec AO', $\pm$ cosec BO'.

We leave the direct demonstration of the foregoing results as an exercise to the student.

METHOD OF RECIPROCATION ON THE SPHERE.

140. From what has been said in Art. 137, it will be seen that the method of reciprocation already explained (see Arts. 42 and 43) may be extended to the sphere. But unless the lesser circle, with respect to which the reciprocation is performed, satisfy a certain condition (to be presently indicated), the method is defective in its spherical applications, chiefly because the angle under two great circles in one figure has no direct representative in the reciprocal figure. This angle is not in general equal to the angle subtended at the spherical centre of the lesser circle by the points of which the great circles are the polars; nor is it measured by the arc joining those points. Before explaining the particular condition above referred to, we shall give a few examples of the cases in which the general method may be adopted with advantage.

1°. We have already given, as examples of the method of projection, theorems on the sphere analogous to Pascal's and Brianchon's theorems *in plano* (see Art. 112). The reader will at once perceive that the two spherical theorems referred to are reciprocals exactly in the same sense as the plane theorems themselves.

2°. The anharmonic ratio of four points on a lesser circle is constant (Art. 128); what is the reciprocal property?

Proceeding in a way analogous to that followed in Art. 42, 3°, we find the following:—

" If four fixed great circles touch a given lesser circle, the anharmonic ratio of the four points, where they are cut by a variable fifth great circle touching the lesser circle, is constant."

3°. The following is the reciprocal of the Proposition given in Art. 132, 2°:—

"If three pairs of arcs be drawn from three points in a great circle to touch a given lesser circle, any seventh great circle touching the lesser circle will be cut by the former six in involution." Because (Art. 137, 2° and 5°) *the reciprocal of a pencil in spherical involution is a system of six points in spherical involution.*

4°. "If a spherical quadrilateral be circumscribed to a lesser circle, any two great circles touching the lesser circle, together with four arcs from their point of intersection to the angles of the quadrilateral, make a pencil in involution."

The reciprocal theorem is,—

"If a spherical quadrilateral be inscribed in a lesser circle, any great circle cutting the circle and the four sides is cut in involution."

In order to establish these two theorems, it is only requisite to prove one (no matter which). This the reader will easily do either by projection, or wholly by the principles of spherical geometry.

141. We stated at the commencement of the last Article that, in spherical reciprocation, it is, in most cases, expedient to select a lesser circle satisfying a certain condition. *This condition is that the square of the tangent of the spherical radius should equal* — 1. Recollecting the definitions of pole and polar (with respect to a real circle), given in Art. 134, it follows that, in the present case, *the pole* O (see fig.) *and its polar* O'R lie on different sides of the spherical centre C, *and that* OC *is the complement of* CO.'. In other words, O *is the pole in the usual trigonometrical sense, of the great circle* O'R, and as the angle between two great circles is measured by the arc joining their trigonometrical poles, the defect before noticed in the general method of spherical reciprocation is thus supplied. In the figure above given the lesser circle has not been drawn, because its radius is, in fact, imaginary. However, it is evident, from

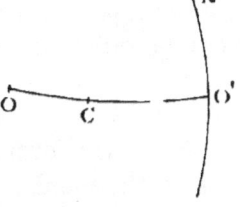

what has been said, that in our applications we may dispense entirely with the lesser circle, so that the circumstance just mentioned will not be productive of any embarrassment.

142. Throughout the remainder of this Chapter we shall employ the word "pole" with its common trigonometrical meaning; and we shall construct the reciprocal of a sperical figure, by taking the poles of the several great circles in the given figure, and joining them by arcs, whose poles will then be corresponding points in the original. This reciprocal relation follows, of course, from the general principle laid down in Prop. 1° of Article 137, which is also evidently true, *a priori*, of poles in the ordinary usage of spherical trigonometry. The well-known relations between a spherical triangle and its *polar or supplemental triangle* are obvious results of the particular mode of reciprocating now under consideration. In what follows, the reader is supposed to be acquainted with the demonstrations of the properties in question.

SUPPLEMEMTARY FIGURES.

143. If we examine the demonstrations referred to in the last Article, it will be perceived that they equally apply to spherical polygons of any number of sides. We shall then assume the two following as elementary and fundamental principles of spherical polygons.

1°. *If the poles of the sides of a spherical polygon be joined by arcs, a new polygon is formed, the poles of whose sides are the corresponding corners of the original.* (Two such polygons may be called *mutually polar.*)

2°. *The angles of either polygon are measured by arcs supplemental to the corresponding sides of the other.*

From the former of the two properties it follows (by supposing the number of the sides to be indefinitely increased) that *the locus of the poles of great circles touching any curve on the sphere is another curve, such that the poles of its tangent arcs are the corresponding points of the original.*

Two such *curves* may be said to be *reciprocally polar*. The term *supplementary* is employed by Chasles to express their mutual relation. It is evident that any great circle perpendicular to one is also perpendicular to the other, and that the intercepted arc is constant, namely, a quadrant.

It is also evident that radii of the sphere drawn to the points of two such curves will form two *cones* (in the widest sense of the term) so related that the tangent planes of each are perpendicular to the *corresponding* sides of the other. Two such cones are said to be *supplementary* to each other.

144 From the preceding properties of spherical polygons some remarkable consequences may be deduced:—

1°. *The angular measure of the area of any spherical polygon, added to the perimeter of its polar polygon, gives a constant sum.*

Assuming the surface of the sphere to be equal to the sum of the areas of four great circles, it can be proved that the area of a spherical polygon of n sides $= r^2 \cdot \{A + B + C + \&c. - (n-2)\pi\}$, r being the radius of the sphere, and A, B, C, &c., the measures of the angles of the polygon, by arcs of a circle whose radius equals 1. The quantity within the vinculum we call the angular measure of the area. Now let a', b', c', &c., be the corresponding sides of the polar or supplemental polygon, and we shall have $A + a' = \pi$, $B + b' = \pi$, &c., and therefore $A + B + C + \&c. - (n-2)\pi + a' + b' + c' + \&c. = n\pi - (n-2)\pi = 2\pi = $ constant. Q. E. D.

If $r = 1$, we may say that the area of the polygon, added to the perimeter of its polar, equals 2π.

2°. *If either polygon be inscribable in a lesser circle, the other is circumscribable to another lesser circle having the same spherical centre as the former, and a complemental spherical radius.*

We may suppose the case of two triangles ABC, A'B'C', but the reasoning will apply to any polygons. Let O be the spherical centre of the lesser circle round the triangle ABC, and let the arcs AO, BO, CO be produced to meet the sides of the polar triangle in the points P, Q, R. Then, as AO, BO, CO are equal, their complements OP, OQ, OR are equal, and therefore O is the spherical centre of a circle inscribed in the triangle A'B'C'. The radii are evidently complemental.

145. *Two curves, polar reciprocals on the sphere* (see Art. 143), possess the following properties:—

1°. *The projections of two such curves on any plane touching* the sphere are polar reciprocals in plano *with respect to a circle whose centre is the point of contact, and the square of whose radius equals* — 1 (see Arts. 42, 4°, and 99, 2°).

This is evident from Articles 134 and 141.

2°. *If one of the curves be closed* (such as that formed by the intersection of the sphere with a cone whose vertex is at the centre), *so will the other, and the spherical area of either, added to the perimeter of the other, equals* 2π.

This appears from Art. 144, 1°.

3°. *If one of the curves be a lesser circle, the other will also be a lesser circle, with the same spherical centre and a complemental spherical radius.*

This follows at once from Art. 144, 2°. It is also evident, *a priori*, that if two circles be constructed having the relations above-mentioned, each of them is the locus of the poles of great circles touching the other.*

Hence it appears that *the cone supplementary* (see Art. 143) *to a right cone is also right.*

146. The reader may perhaps be glad to see the second Proposition of the preceding Article verified in the case of two lesser circles mutually polar.

Let AS be an infinitely small portion of a circle, by whose revolution on its diameter CT (see fig.), the sphere is supposed to be generated. Let AB be perpendicular to CT, and SO to AB. The band on the surface generated by $AS = AS \cdot$ (the circumference of a circle whose radius is AB) $= AS \cdot 2\pi \cdot AB$. Now, considering ASO as a triangle, it is similar to CAB, and therefore $\dfrac{AS}{SO} = \dfrac{CA}{AB}$, and

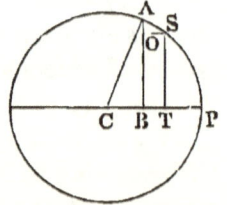

consequently $AS \cdot AB = CA \cdot SO = CA \cdot BT$. The band before mentioned is therefore equal to $2\pi \cdot CA \cdot BT$, that is, *the portion of the spherical surface generated by an infinitesimal arc* AS *equals the circumference of the generating circle multiplied*

---

* Since every great circle has two poles, the locus of the poles of tangent arcs to a spherical curve consists of two curves situated on opposite parts of the sphere, and equal in all respects. As, however, one of those curves will always possess the properties laid down as belonging to the *reciprocal* of the given curve, to that one this denomination may be confined.

by the part of the diameter, BT, *intercepted by perpendiculars from the extremities of the arc*. Hence it immediately follows that *the same result holds good for any finite arc* of the generating circle, for instance, AP. Taking the radius of the sphere $= 1$, we have then for the spherical area of the lesser circle generated by the arc AP, $2\pi \cdot$ BP, or 2π $(1 - \cos \text{AP})$. But the perimeter of its polar circle $= 2\pi \cdot$ (sine of the complement of AP) $= 2\pi \cdot \cos$ AP. Therefore the spherical area of the one circle, added to the perimeter of the other, equals 2π. Q. E. D.

It is evident from the foregoing demonstration that the surface of the sphere $= 2\pi \cdot \text{CA} \cdot 2\text{CP} \cdot = 4\pi \cdot \text{CP}^2 =$ four times the area of a great circle.

As this result does not presuppose any principle of spherical areas (not here proved), it may be made the basis of the investigation of the area of a spherical polygon to be found in trigonometrical works.

147. Let us now conceive a spherical polygon whose sides are any curves *not great circles*, and it will be seen that *its reciprocal is another polygon formed by curves, the reciprocals of the former, and by arcs of great circles supplemental to the angles of the original*. It will also be seen that each of these supplemental arcs *touch* at its extremities the two adjacent curves. Should any of the sides of the original polygon become a great circle, its reciprocal curve vanishes into a point. In all cases the property laid down in Art. 114, 1°, holds good; and therefore *questions concerning the areas of any figure on the sphere are reducible to others regarding corresponding perimeters* and vice versâ.

EXAMPLES ON SPHERICAL RECIPROCATION.

148. The theory of reciprocal curves on the sphere is also useful in effecting the mutual reduction of questions of another class. Suppose a great circle to move on the sphere so as to satisfy a certain condition, and let it be required to find the fixed curve which it will always touch (that is, its envelope). Now if the given condition enable us to find the locus of the pole of the great circle, it is evident, from Art. 143, that the reciprocal of the locus will determine the envelope required. *The finding of envelopes is*, in

U

this way, *reduced to the finding of corresponding loci,* and *vice versâ.** The following examples will illustrate what has been said:—

1°. *Given in position the vertical angle of a spherical triangle; given also the sum of the cosines of its base angles; required the envelope of the base.*

By means of the polar triangle, the question proposed is reduced to the following:—

"Given in magnitude and position the base of a spherical triangle; given also the sum of the cosines of its sides; to find the locus of the vertex."

Let a and b be the sides, and c the base; let ϕ represent the arc joining the vertex to the middle of the base, and θ the angle made by the joining arc with the base. We have then

$$\cos a = \cos \phi \cdot \cos \tfrac{1}{2} c + \sin \phi \cdot \sin \tfrac{1}{2} c \cdot \cos \theta,$$
and
$$\cos b = \cos \phi \cdot \cos \tfrac{1}{2} c - \sin \phi \cdot \sin \tfrac{1}{2} c \cdot \cos \theta,$$
whence,
$$\cos a + \cos b = 2 \cos \phi \cdot \cos \tfrac{1}{2} c.$$

This equation determines ϕ, and the locus is therefore (in general) a *lesser* circle, whose spherical centre is the middle of the base.

When the given value of $\cos a + \cos b$ is zero, $\phi = 90°$, and the locus becomes a *great* circle.

Returning now to the triangle in the original question, it follows from Art. 145, 3°, that the envelope of the base is a lesser circle, which vanishes into a point when the given sum of the cosines of the base angles equals zero.

2°. *Given in magnitude and position the base of a spherical triangle; given also its area; required the locus of its vertex.*

Let AB be the base of its spherical triangle, and C its vertex; then, if A'B'C' (see fig. on next page) be the polar triangle, we shall have its vertical angle C' given in position, and its perimeter also given. Now, if a circle be *exscribed* (see Art. 6, 4°) to A'B'C', touching its base, the perimeter of A'B'C' is evidently C'O + C'O'

---

* If a great circle touch any curve on the sphere, it will also touch another curve equal to the former in all respects, and placed on the opposite part of the sphere. As, however, one is determined by the other, we shall, for the sake of simplicity, consider a spherical envelope as a single curve.

(O and O' being the points of contact with C'A' and C'B'), or, $2C'O$; therefore the points O, O' are given, and consequently the exscribed circle itself is fixed, which circle is therefore the envelope of the base A'B' of the polar triangle. The locus of C is (Art. 145, 3°) either the lesser circle reciprocal to the exscribed circle, or another lesser circle equal to this reciprocal, but on the opposite part of the sphere. It is evident from the figure that, in the present case, the latter lesser circle is the one to be taken.

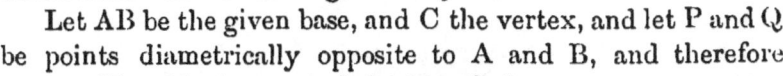

As this question is one of importance, we shall add the following direct investigation taken from Thomson's Trigonometry:—

Let AB be the given base, and C the vertex, and let P and Q be points diametrically opposite to A and B, and therefore given. Now *if a lesser circle be described round the triangle PQC, it is easy to prove that this circle remains constant, and is therefore the locus required.* For, as the area of ABC is given, the sum of its angles is given; but this sum $= \text{AQC} + \text{BPC} + \text{PCQ} = \text{AQC} + \text{BPC} + \text{CQR} + \text{CPR}$ (R being the spherical centre of the lesser circle) $= \text{AQR} + \text{BPR} = 2 \cdot \text{AQR}$ (since the angles RQP and RPQ are equal); therefore the angles AQR and BPR are given. It follows at once that the circle is fixed.

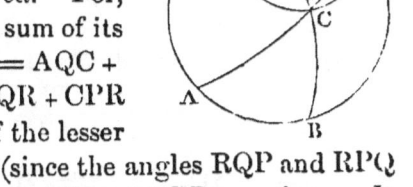

3°. *Given the vertical angle of a spherical triangle, and the ratio of the cosines of its base angles, to find the envelope of the base.*

Let us conceive the polar triangle as before, and the question is reduced to this:—

Given in magnitude and position the base of a spherical triangle; given also the ratio of the cosines of its sides; find the locus of its vertex.

Let the sides of the triangle be a, b, and let x, y be the segments of the base made by a perpendicular p from the vertex. We have then, $\cos a = \cos p \cdot \cos x$, and $\cos b = \cos p \cdot \cos y$, therefore $\dfrac{\cos a}{\cos b} = \dfrac{\cos x}{\cos y}$, that is, the cosines of the sides are to one another as the cosines of the segments made by the perpendicular.

As $\dfrac{\cos x}{\cos y}$ is a given quantity, $\dfrac{\cos y - \cos x}{\cos y - \cos x}$ or, $\tan \tfrac{1}{2} (x + y)$ $\tan \tfrac{1}{2} (x - y)$ is also given. If this quantity is less than $\tan^2 \tfrac{1}{2}$ base, the perpendicular falls within the triangle; and, if greater, without. In either case x and y are determined, and *the locus required is therefore a great circle perpendicular to the base.*

It follows from this that *the base of the triangle, in the original question, passes through a fixed point.* This fixed point, which may be considered as an infinitely small circle, is the envelope required.

4°. Our next example requires the following preliminary Proposition:—

LEMMA 16.—*Through a given point* O (see fig.) *on the surface of the sphere, let any arc,* PQ, *be drawn cutting a given lesser circle, whose spherical centre is* C, *the product of the tangents of the spherical semi-angles subtended at* C *by the segments of the arc is constant.*

Let CM be perpendicular to PQ; then

$$\cos OCM = \dfrac{\tan CM}{\tan OC},$$

and $\cos QCM = \dfrac{\tan CM}{\tan CQ}$; therefore $\dfrac{\cos OCM}{\cos QCM} = \dfrac{\tan CQ}{\tan OC}$,

and $\dfrac{\cos OCM - \cos QCM}{\cos OCM + \cos QCM} = \dfrac{\tan CQ - \tan OC}{\tan CQ + \tan OC}$,

or,

$\tan \tfrac{1}{2} OCQ \cdot \tan \tfrac{1}{2} OCP = \dfrac{\tan QC - \tan OC}{\tan QC + \tan OC} = $ constant. Q. E. D.

It follows at once conversely that if $\tan \tfrac{1}{2} OCQ \cdot \tan \tfrac{1}{2} OCP$ be given, and also the circle, the distance CO is given.

Suppose now the following question to be proposed:—

Given in position a lesser circle and a great circle touching it; required the locus of the intersection of two other great circles also touching it, and intercepting a quadrant on the fixed great circle.

Taking, as before, the reciprocal circle and the polar triangle, the question is immediately reduced to this:—

If a right-angled spherical triangle be inscribed in a given lesser circle its vertex being at a fixed point, find the envelope of its hypotenuse.

Let R be the fixed vertex of the right-angled triangle PQR (see last figure) inscribed in the given lesser circle, whose spherical centre is C, and let the arc RC be produced to meet PQ in the point O. Draw CS and CT perpendicular to the sides of the right-angled triangle. We shall then have

$$\cos CR = \cot SCR \cdot \cot SRC, \text{ and } \cos CR = \cot TCR \cdot \cot TRC,$$

and therefore $\cos^2 CR = \cot SCR \cdot \cot TCR \cdot \cot SRC \cdot \cot TRC =$
$$\cot\tfrac{1}{2} PCR \cdot \cot\tfrac{1}{2} QCR = \tan\tfrac{1}{2} PCO \cdot \tan\tfrac{1}{2} QCO.$$

Since CR is a given quantity, it follows from the foregoing Lemma, that the distance CO is constant, and therefore the point O is fixed. The envelope of the hypotenuse PQ is therefore this fixed point.

The locus required in the question first proposed is, of course, a great circle.

149. The following examples will serve still further to illustrate the mutual relation of certain questions, both in plane and spherical geometry:—

1°. The product of the tangents of the halves of the segments of an arc through a given point, and cutting a given lesser circle, is constant (Art. 133, Lemma 12); what is the reciprocal theorem?

Recollecting that if any number of poles are on a great circle, the great circles, whose poles they are, pass through one point, and that the arc joining two poles measures the angle made by the corresponding circles, we find the following:—

If tangent arcs be drawn from any point in a given great circle to a given lesser circle, the product of the tangents of the semi-angles made by them with the great circle, is constant, the angles being measured in opposite directions with respect to the given great circle.

The theorem just given, being independent of the radius of the sphere, will remain true when the radius becomes infinite; we have then the analogous theorem in plane trigonometry:—

"If from any point in a given right line tangents be drawn to a given circle, the product of the tangents of the semi-angles made by them with the opposite portions of the right line is constant."

If again, this last result be reciprocated by the method already explained in Art. 42, we shall have a theorem *in plano* analogous to the Lemma given in Art. 148, 4°.

2°. It is easy to prove that *if a spherical quadrilateral be inscribed in a lesser circle, the product of the sines of the halves of the diagonals is equal to the sum of the products of the sines of the halves of the opposite sides.* In fact, we need only draw the chords of the diagonals and sides of the spherical quadrilateral. We shall then have a plane quadrilateral inscribed in a circle, in which, by a well-known Proposition (Lardner's Euclid, B. vi. Prop. 17), the rectangle under the diagonals equals the sum of the rectangles under the opposite sides, and as the chord of an arc equals twice the sine of half the arc, the theorem enunciated becomes evident.

This Proposition being established, we have the following reciprocal:—

"If a spherical quadrilateral be circumscribed to a lesser circle, the product of the cosines of the halves of the internal angles, made by the opposite sides produced, is equal to the sum of the products of the cosines of the halves of the opposite angles of the quadrilateral."

By supposing the radius of the sphere to become infinite, we shall have a theorem *in plano*, the reciprocal of the geometrical theorem above cited.

3°. We have already proved (Art. 125) that if through one point three arcs be drawn from the angles of a spherical triangle, meeting the opposite sides, the products of the sines of the alternate segments so made are equal. The reciprocal theorem is as follows:—

If a great circle be drawn cutting the three sides of a spherical triangle, and each point of intersection be joined to the opposite angle, the products of the sines of the alternate segments of the angles of the triangle made by the joining arcs are equal.

By making the radius of the sphere infinite as before, we have a corresponding Proposition in plane trigonometry.

If this Proposition, also, be reciprocated, we find the following:—

"If through one point lines be drawn from the angles of a plane triangle meeting the opposite sides, the product of the sines of the angles subtended at any point in the plane of the triangle by the alternate segments of the sides are equal." It is evident, however, from Article 110, that this last result is merely equivalent to the known relation between the segments themselves.

4°. LEMMA 17.—*The perpendiculars from the angles of a spherical triangle upon the opposite sides meet in a point.*

Let CP be one of the perpendiculars of the spherical triangle ABC. Then, since $\sin AP = \tan CP \cdot \cot A$, it follows that the products of the sines of the alternate segments of the sides made by the three perpendiculars are equal, each product being equal to the continued product of the tangents of the three perpendiculars, and the cotangents of the three angles of the triangle. Hence, by Art. 125, Lemma 10, the perpendiculars meet in a point.

This Lemma being laid down, we shall now prove the following properties of two spherical triangles *mutually supplemental*.

"If the corresponding corners of two such triangles be joined, the joining arcs meet in a point."

For, the joining arcs are evidently the perpendiculars from the angles of either triangle on its opposite sides.

"If the corresponding sides of the two triangles be produced to meet, their intersections lie on a great circle."

This follows from the last, being, in fact, its reciprocal.*

---

* The two properties here given are only particular cases of the following:—
If two spherical triangles be mutually polar with respect to a lesser circle (that is, such that every side in each is the polar of a corresponding vertex in the other), *the points of intersection of corresponding sides are on the same great circle; and the arcs joining corresponding angles meet in a point.*

This Proposition corresponds to that contained in Art. 51, and may be proved on similar principles, or deduced from the latter by projection.

If we suppose the square of the tangent of the spherical radius of the lesser circle to be equal to -1, we have (Art. 141) the properties given in the text.

CHAPTER X.

RADICAL AXIS AND CENTRES OF SIMILITUDE ON THE SPHERE.

150. *To find the locus of a point on the sphere from which arcs drawn to touch two given lesser circles are equal.*

Conceive a figure on the sphere analogous to that in Art. 53.

Then, $\dfrac{\cos AO}{\cos TA} = \cos OT$, and $\dfrac{\cos BO}{\cos BT'} = \cos OT'$;

therefore, $\dfrac{\cos AO}{\cos BO} = \dfrac{\cos AT}{\cos BT'}$.

We have now the base AB of a spherical triangle ABO, and the ratio of the cosines of its sides, to find the locus of the vertex. The locus required is therefore (Art. 148, 3°) a great circle OP, perpendicular to the arc joining the spherical centres A and B, and cutting it so that $\dfrac{\cos AP}{\cos BP} = \dfrac{\cos AT}{\cos BT'}$.

This great circle may be called *the radical axis* of the two given lesser circles. It coincides with the arc joining the points of intersection, in case the circles cut one another, and with the tangent arc drawn at the point of contact in case they touch. *In all cases the products of the tangents of the halves of the segments of arcs drawn from any of its points cutting the given circles respectively, are equal* (the segments being measured from the point) (Art. 133, Lemma 12).

151. By reasoning analogous to that contained in Art. 54, it will be found that in all cases there exists an infinite system of circles on the sphere, having as a common radical axis the arc OP, and having their spherical centres along the arc AB. We have in fact only to substitute for $BP^2 - BT'^2$ and $AP^2 - AT^2$ respectively, the fractions $\dfrac{\cos BP}{\cos BT'}$ and $\dfrac{\cos AP}{\cos AT}$.

It will also be found that when the two given circles do not meet one another, there will be two limiting points such as F and

F′, at equal distances on each side of P, determined by the equation $\cos FP = \dfrac{\cos AP}{\cos AT}$. These limiting points, as before, are to be regarded as infinitely small circles belonging to the system.

All the results arrived at in Art. 55 hold good on the sphere; the words centre and radius being understood, of course, to mean spherical centre and spherical radius.

152. "If a system of lesser circles have a common *ideal* chord, and a great circle be drawn cutting them respectively in the points AA′, BB′, CC′, and so on, and cutting the chord in O, the middle points of the arcs OA, OA′, OB, OB′, &c., will form a system of points in spherical involution, of which the point O is one of the centres."

This is evident from what has been said in Arts. 133 and 150.

It is also evident that the foci are in this case always real, and are determined by measuring on the great circle a portion on each side of O, equal to the half of the arc joining O to either of the limiting points.

THE RADICAL CENTRE ON THE SPHERE.

153. "The radical axes belonging to every pair of three given lesser circles on the sphere meet in a point, which is either within all three circles or without them all." (This point is called the *radical centre* of the three circles.)

This follows by reasoning analogous to that contained in Art. 62.

We have also the following Proposition on the sphere:—

"If any number of lesser circles pass through the same two points, and intersect a given lesser circle, the *spherical chords* of intersection all pass through one point lying on the great circle joining the two points."

When the radical centre of three lesser circles is without them all, it is evidently the spherical centre of a circle cutting them orthogonally.

154. "Let two spherical triangles be constructed, having for sides the spherical chords (real or ideal) determined by the intersection of three given lesser circles with two others respec-

tively; it is required to prove that the points of intersection of the corresponding sides lie on the same great circle."

Proceeding as in Art. 63, we readily see that the theorem to be proved is reduced to the following:—

"If two spherical triangles be such that arcs joining corresponding angles meet in a point, the points of intersection of corresponding sides will lie on a great circle." This follows at once from Art. 23 by projection; and therefore the theorem proposed is proved.

155. We shall conclude this subject with the following problem:—

"To describe a circle such that the radical axes determined by it and three given lesser circles shall pass respectively through three given points on the sphere."

This problem corresponds to that given in Art. 64, and is readily reduced to the following:—

To describe a spherical triangle having its angles on three given great circles meeting in a point, and its sides passing each through a given point. Now, one side of the spherical triangle is fixed completely (and therefore the problem solved) by means of the spherical Proposition which corresponds to the theorem proved in Art. 22. That Proposition is as follows:—"If the angles of a spherical triangle move on three given great circles meeting in a point, and if a definite pair of its sides pass through two given points respectively, the third side also constantly passes through a fixed point on the great circle joining the two given points." The truth of this spherical theorem is evident, from Art. 22, by projection; and the fixed point may be found by taking a random position of the triangle. The solution of the original problem is now evident from what has been said.

It is easy to see that when the three given points are on the same great circle, the problem is either impossible or indeterminate.

CENTRES OF SIMILITUDE OF TWO LESSER CIRCLES.

156. If the arc joining the spherical centres of two lesser circles be divided internally and externally, so that the *sines* of the segments are in the ratio of the sines of the corresponding spheri-

cal radii, the points of section possess certain properties analogous to those of the centres of similitude of two circles *in plano*, and may be called by the same name.

1°. A great circle touching the two given circles (if such be possible) passes through one of the centres of similitude.

For, let us conceive a spherical figure analogous to that in Art. 65, and let TT′, the common tangent, cut the arc joining the spherical centres in C; we have then

$$\frac{\sin AT}{\sin AC} = \sin ACT = \frac{\sin BT'}{\sin BC},$$

that is, sin AT : sin BT′ :: sin AC : sin BC, and therefore the point C is the external centre of similitude. If the tangent were transverse, it would pass through the internal centre.

2°. *If a great circle be drawn from either centre of similitude, cutting the given circles, the tangents of the halves of the arcs between this centre and a pair of corresponding points are in a constant ratio.*

Let us take the external centre of similitude, for instance, and let (see fig.) the arc OO′, between a pair of corresponding points, be bisected in M, and let the arcs AM and BO′ be drawn, and produced to meet. We

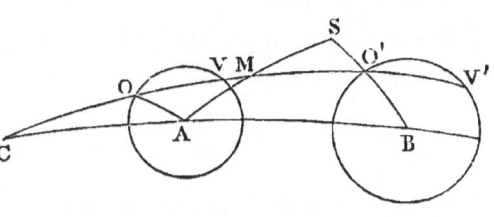

have then $\dfrac{\sin CA}{\sin AO} = \dfrac{\sin COA}{\sin ACO}$ and $\dfrac{\sin CB}{\sin BO'} = \dfrac{\sin CO'B}{\sin BCO'},$

and therefore sin COA = sin CO′B, and consequently, by the definition of corresponding points (see Art. 65, 2°), the angles COA and CO′B are equal. It follows that the triangles AOM and SO′M are equal in all respects, and therefore that O′S (being equal to AO) is of constant length. Now, considering AS as a transversal arc cutting the three sides of the spherical triangle BCO′, we have (Art. 125)

sin BS · sin CA · sin MO′ = sin SO′ · sin BA · sin CM,

and therefore $\dfrac{\sin CM}{\sin MO'} = \dfrac{\sin BS \cdot \sin CA}{\sin SO' \cdot \sin BA} = $ constant.

Hence, $\dfrac{\sin CM + \sin MO'}{\sin CM - \sin MO'} = \dfrac{\tan\frac{1}{2}(CM + MO')}{\tan\frac{1}{2}(CM - MO')} =$ constant,

and finally $\dfrac{\tan\frac{1}{2} CO'}{\tan\frac{1}{2} CO} =$ constant. *Q. E. D.*

A similar proof will apply to the case of the internal centre.

3°. *The product of the tangents of the halves of the arcs between either centre of similitude and a pair of points inversely corresponding* (see Art. 65, 3°), *is constant.*

For, $\dfrac{\tan\frac{1}{2} CV \cdot \tan\frac{1}{2} CO'}{\tan\frac{1}{2} CV \cdot \tan\frac{1}{2} CO} = \dfrac{\tan\frac{1}{2} CO'}{\tan\frac{1}{2} CO} =$ constant;

but (Art. 133, Lemma 12) $\tan\frac{1}{2} CV \cdot \tan\frac{1}{2} CO$ is constant; therefore $\tan\frac{1}{2} CV \cdot \tan\frac{1}{2} CO'$ is constant. This proves the Proposition in the case of the external centre, and a similar proof will apply in the other case.

4°. "If a lesser circle touch two others, the arc joining the points of contact passes through one of their centres of similitude."

Let us conceive a spherical figure analogous to the first figure in Art. 65, 4°, and we shall have the angle $XO'V = AOV$, and therefore the angle $BO'C = AOC$, and consequently

$$\dfrac{\sin BC}{\sin BO'} = \dfrac{\sin AC}{\sin AO}.$$

This proves the Proposition in the case when the two contacts are of the same kind, and a similar proof will hold when they are of different kinds.

The points of contact are (as *in plano*) points inversely corresponding.

157. "If an infinite system of lesser circles touch two given ones, the radical axes of every pair that can be taken pass through the external centre of similitude when the two contacts are of the same kind, and through the internal centre when of different kinds. In the former case, a definite circle can be found which cuts orthogonally all the circles of the system, and in the latter case this circle becomes imaginary."

These theorems are analogous to those given in Art. 66, 2°,

and are proved in a similar manner by means of Art. 156, 4°, and Lemma 12 of Art. 133.

158. The theorems stated in Art. 157 lead to the following spherical Propositions analogous to the results arrived at in Arts. 89 and 90:—

1°. "If two lesser circles touch one another, and also touch two given lesser circles, either both externally or both internally, the locus of their point of contact is a circle having for its spherical centre the external centre of similitude of the two given circles."

2°. "The envelope of a variable lesser circle, which touches one given lesser circle, and cuts another orthogonally, is a lesser circle."

159. The following results are analogous to some of those contained in Art. 87:—

1°. "If from either centre of similitude of two lesser circles two great circles be drawn cutting them, a pair of spherical chords, inversely corresponding, intersect on the radical axis of the two circles."

For, if we conceive the figure in the Article referred to, to be adapted to the sphere, we shall have (Art. 156, 3°)

$$\tan\tfrac{1}{2} CV \cdot \tan\tfrac{1}{2} CO' = \tan\tfrac{1}{2} CP' \cdot \tan\tfrac{1}{2} CQ,$$

and therefore the four points V, P', Q, O' lie on a lesser circle (Art. 133, Lemma 12), and consequently

$$\tan\tfrac{1}{2} XO' \cdot \tan\tfrac{1}{2} XP' = \tan\tfrac{1}{2} XV \cdot \tan\tfrac{1}{2} XQ.$$

It follows from this (Art. 150) that X lies on the radical axis.

2°. "Great circles touching the lesser circles at a pair of points inversely corresponding also intersect on the radical axis, and consequently make equal angles with the great circle joining the points."

3°. "Great circles touching the lesser circles at a pair of corresponding points make, with the great circle joining the points, equal alternate angles."

AXES OF SIMILITUDE ON THE SPHERE.

160. *The six centres of similitude of three lesser circles taken in pairs lie, three by three, on four great circles, which may be called axes of similitude.*

This Proposition corresponds to that contained in Art. 67, and may be proved in a similar manner, by conceiving the figure in that Article to be adapted to the sphere, and by introducing the sines of the various arcs corresponding to the right lines made use of in the former proof. Thus, in place of the equation

$$\frac{BC'}{AC'} = \frac{R'}{R}, \text{ we shall have } \frac{\sin BC'}{\sin AC'} = \frac{\sin R'}{\sin R},$$

and so on. The proof is completed by the aid of Lemma 11 of Art. 125.

We have also the following Proposition, analogous to that given in Art. 94, 1°:—

"In a system of three lesser circles on the sphere, the arcs joining each of the spherical centres to the internal centre of similitude of the other two circles, meet in a point."

161. *Given three lesser circles on the sphere, four pairs of circles may be described touching them, by the aid of the four axes of similitude,* in a manner analogous to that explained in Art. 68, with respect to circles *in plano*. The reader will find no difficulty in establishing the following results, which correspond exactly to those already given in Arts. 68 and 94.

1°. Each axis of similitude is the radical axis of a corresponding pair of circles, touching the three given ones in such a manner that one set of contacts are respectively opposite in kind to the other set.

2°. The radical centre of the three given circles is the internal centre of similitude of the pair in question.

3°. The arcs joining the radical centre to the poles of the axis of similitude, with respect to the three given lesser circles, coincide with the three spherical chords joining the three pairs of contacts respectively.

4°. A great circle drawn from the radical centre perpendicular to the axis of similitude, passes through the spherical centres of the corresponding pair of touching circles.

5°. The circle orthogonal to the three given circles, belongs to each of the systems of circles, having radical axis in common with the four pairs of touching circles respectively.

CHAPTER XI.

PROPERTIES OF THE SPHERE CONSIDERED IN RELATION TO SPACE.

162. THE properties of the sphere which have hitherto engaged our attention are principally such as belong to its *surface*. We now propose to explain some of its important properties when considered in relation to *space*. We shall commence with those concerning a point and a plane mutually polar. The full importance of these properties can only be seen in connexion with the theory of surfaces of the second degree; but the elementary principles themselves may, with propriety, be introduced into a work of the present nature.

"Given a point O, and a sphere whose centre is C; let CO be joined, and on the joining line a portion CO′ be taken (in the same direction from C as O is), such that $CO \cdot CO' = $ the square of the radius of the sphere, a plane perpendicular to the line CO′, at the point O′, is called the *polar plane* of the point O, with respect to the given sphere, and the point O is called the *pole* of the plane."

From these definitions it is evident that when the pole is within the sphere, its polar plane is without it; and when the pole is on the sphere, its polar plane touches the sphere at the pole.

When the pole is without the sphere, it appears, from Art. 38, that the polar plane cuts the sphere in a circle which is the curve of contact of a right cone circumscribed to the sphere, and having its vertex at the pole.

If the radius of the sphere were to vary, the point O and the centre remaining unchanged, we should have an infinity of such curves of contact, which would all lie on a sphere having CO as diameter (Euclid, B. iii. Prop. 31).

163. *Any right line through the pole is cut harmonically by the polar plane and the sphere.*

For, draw through the right line and the centre a plane. This plane will cut the sphere in a circle, and the polar plane in a right line, which will be (Art. 38) the polar of O with respect to the circle. The Proposition is then evident from Art. 39.*

164. The preceding Proposition leads to results of great importance from their application to the theory of surfaces.

1°. *If any number of points on a plane be taken as poles, their polar planes, with respect to a given sphere, pass through one point, namely, the pole of the plane.*

For, let O be the pole of the given plane, and R any point on it. Then, since the line joining O and R is (by last Art.) cut harmonically in real or imaginary points, it is evident (by the same Art.) that the polar plane of R must pass through O.

Hence it follows that *if a number of cones be circumscribed to a sphere, with their vertices all on the same plane, the planes of their bases pass through one point.*

2°. *If any number of planes pass through a point, the locus of their poles with respect to a given sphere is a plane, namely, the polar plane of the point.*

For, let O be the given point, and conceive any plane to be drawn through it, and the pole of the plane to be joined to O; then, as the joining line is cut harmonically, the pole of the plane drawn through O must lie on the polar plane of O.

Hence *if a number of cones be circumscribed to a sphere, with the planes of their bases passing through one point, the locus of their vertices is a plane.*

Propositions 1° and 2° may also be proved in a manner similar to that explained in the Note to Art. 40.

165. The following Propositions are also of importance:—

---

* It follows from this Proposition that a plane through any point cuts the polar plane of the point in a right line, which is the polar of the same point with respect to the circular section of the sphere made by the plane. When the plane does not meet the sphere the circle becomes imaginary, while the pole and polar, with respect to it, remain real. This is an example of the case supposed in Art. 99, 2°.

1°. *If any number of points lie on a right line, their polar planes with respect to a sphere pass through another right line.*

Draw a plane through the centre and the given line. The polar planes are all perpendicular to this plane, and their intersections with it are (Art. 38) the polars of the points on the right line with respect to the circular section of the sphere made by the plane. These intersections pass through one point (Art. 40, 1°), namely, the pole of the given line with respect to the section above mentioned. It follows that the polar planes all pass through a right line drawn through this pole perpendicular to the plane of the section.

It is easy to see that the given right line, and that through which the polar planes all pass, are mutually convertible.

2°. *If any number of planes pass through one right line, the locus of their poles is another right line.*

Draw a plane through the centre perpendicular to the given right line. This plane will contain all the poles whose locus we are seeking, and (Art. 38) will cut the polar planes in a number of right lines, the polars of the points belonging to the locus with respect to the circular section of the sphere made by the drawn plane. Since all these polars pass through one point, the required locus is (Art. 40, 2°) a right line.

It is evident that the line thus determined and the given line have exactly the same relation to each other as the two corresponding lines in the last Proposition. Two right lines so related are said to be each the *polar line* of the other with respect to the sphere.

3°. *If four points taken as poles lie on one right line, their anharmonic ratio is the same as that of their four polar planes.*

In the first place it appears from Prop. 1° that the four polar planes intersect in a right line, and therefore form a system whose anharmonic ratio (Art. 116) depends upon the mutual inclinations of the planes. Again, the pencil formed by lines joining the centre to the four poles consists of angles respectively equal to those contained by the four planes. Therefore, the anharmonic ratio of the four planes is the same as that of the pencil, and therefore the same as that of the four poles.

Y

RECIPROCAL SURFACES.

166. We shall now explain the general relation existing between *two surfaces, polar reciprocals,* with respect to a sphere.

Let a tangent plane be drawn at any point A' (see figure in Art. 42, 4°) of a given surface of any kind (which the figure can only represent in section), and let the pole A of the tangent plane A'P' be taken with respect to a given sphere, whose centre is C, the locus of the poles of the tangent planes found by varying the point A' on the given surface, will be, in general, a new surface, between which and the original a remarkable reciprocity exists.

In order to see this reciprocity, let us conceive three positions of the point A'; this will give us three tangent planes intersecting in a point which we shall call K, and three corresponding positions of the point A, which will lie in a plane, the polar plane of the point K (Art. 164, 2°). Now let the three positions of A' approach until they coincide; the point K will then coincide with A', and the three corresponding positions of A will also become coincident, and their plane will become the tangent plane to the new surface at the point A. The point A' is therefore the pole of this tangent plane; that is to say, *the point of contact of each tangent plane of the original surface is the pole of the tangent plane at the corresponding point of the new surface*, and therefore the original surface admits of being generated from the new one exactly in the same way as the latter was generated from it. In this the reciprocity consists.

It is obvious, from what has been said, that CA' is perpendicular to the tangent plane AP, and CA to A'P', and that $CA \cdot CP' = CA' \cdot CP =$ the square of radius of the sphere.

167. As an example on the preceding Article, let it be required *to find the surface reciprocal to a given sphere with respect to another given sphere.*

Let S be the centre (see figure on next page) of the given sphere, whose reciprocal we want to find, and let its radius be called r. Let C be the centre of the given sphere, with respect to which the reciprocation is to be performed, and let its radius be called k. (This sphere is not required to be drawn on the figure.)

Let CP be a perpendicular from C, on any plane touching the former sphere at A, and let CA' be taken on it, such that $CP \cdot CA' = k^2$; we have to find the surface, which is the locus of the point A'. Join SA, and draw SD perpendicular on CP. Now, since CP and SA are perpendicular to the tangent plane, they are parallel to one another, and

$$DP = SA = r;$$

also $\quad CD = CS \cdot \cos DCS;$

and therefore $CP = r + d \cdot \cos \theta$ (CS being called d, and the angle DCS, θ). Finally, calling ρ the radius vector CA', we have

$$\rho (r + d \cdot \cos \theta) = k^2$$

or
$$\rho = \frac{k^2}{r + d \cdot \cos \theta}$$

for the polar equation of the locus sought. By comparing this with the known polar equation of a curve of the second degree, referred to one of the foci, it will be seen that the reciprocal surface required is that generated by the revolution of such a curve on its major axis. The generating curve, in the present case, has for one of its foci the given point C, and its axis major coincident with the line CS; its semi-parameter $= \frac{k^2}{r}$, and its excentricity $= \frac{d}{r}$. *The locus is therefore a prolate spheroid when the point C is within the sphere whose centre is S. The spheroid changes into a paraboloid when C is on the sphere, and becomes a hyperboloid when it is without.*

If the question were "to find the curve reciprocal to a given circle, whose centre is S (see last figure), with respect to another given circle whose centre is C," the process shows that the result would be an ellipse, hyperbola, or parabola, according as C is within the former circle, without it, or on its circumference; and the remarks above made, with respect to the focus, axis major, semi-parameter, and excentricity of the curve generating the reciprocal surface, would apply to the reciprocal curve.

This reciprocal curve possesses also the following property:—

Its directrix, corresponding to the focus C, *is the polar of* S *with respect to the circle whose centre is* C. For, from the fundamental property of the directrix, it follows that the distance from C to the corresponding directrix, equals the semi-parameter divided by the excentricity, $= \dfrac{k^2}{d}$, which proves the Proposition.

The reader will find a geometrical investigation of the reciprocal of a circle, in Art. 298 of Salmon's Conic Sections, in which work the theory of reciprocal curves is developed with perfect generality, and its application illustrated by a variety of examples.

168. It has been already intimated, that by the method of reciprocation every property becomes double, and that, in consequence, our means of discovery become enlarged. We shall give here an example connected with the subject of this Chapter.

"A cone circumscribed to a sphere is right, and its curve of contact is a circle." What are the reciprocal results?

If we take the reciprocal of the sphere with respect to another sphere, whose centre we shall call C, we shall have a prolate surface of revolution of the second degree, with C for one of its foci (Art. 167). Also, if we drop perpendiculars from C upon the tangent planes of the cone, these perpendiculars will form another cone, which will be *right* (Art. 145), as being similar to the cone supplementary to the former. Again, as the tangent planes of the cone, circumscribed to the sphere, also touch the sphere, these perpendiculars will pierce the new surface in a series of points lying in one plane (Arts. 166 and 164, 2°). Moreover, planes touching the latter surface at these points will all meet in one point (Arts. 166 and 164, 1°), as being polar planes of the points along the circle of contact of the original cone. Finally, the line joining this point to C makes, with any of the perpendiculars above mentioned, a constant angle, being equal to the inclination of any plane touching the original cone to the plane of its circle of contact (since the angle made by two planes is equal to that subtended by their poles at C, the centre of the reciprocating sphere). Hence we obtain the following results:—

1°. *If a plane be drawn cutting a prolate surface of revolution* of the second degree, the right lines drawn from one of the foci to the points of the curve of intersection are the sides of a right cone.

2°. *Planes touching the prolate surface at the points of the curve of intersection meet in one point.*

3°. *The right line joining this point to the focus is the axis of the right cone mentioned in* Prop. 1°.

If we conceive a plane to be drawn through the point mentioned in Prop 2°, and through the axis of the surface, we get the following general property of curves of the second degree:—

The line joining one of the foci to the intersection of two tangents bisects the angle subtended at the focus by the points of contact.

This result may be directly arrived at by reciprocating the property, possessed by two tangents to a circle, of making equal angles with the chord of contact. This will readily appear from the latter part of Art. 167 and Art. 42, Prop. 4°.

STEREOGRAPHIC PROJECTION.

169. In the next place we shall explain the fundamental Propositions relating to the Stereographic Projection of the Sphere.

In the projection so called, the centre of projection is not at the centre of the sphere, but at the pole of a great circle, on whose plane a figure on the sphere is projected. The theorems above referred to are as follow:—

1°. *The stereographic projection of a circle* (great or lesser), *not passing through the centre of projection, is a circle.*

Let CC' be the diameter of the sphere perpendicular to the plane of projection, and let P be the pole of the circle to be projected, C being the centre of projection. Let the plane of the great circle through C, P, C', cut the plane of projection in the line AB, and the plane 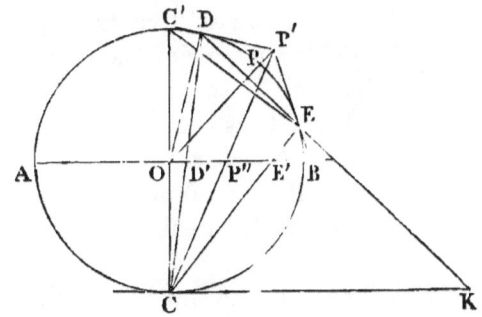 of the circle to be projected in the line DE (see fig.). Now, if we consider the last-mentioned circle as the base of a cone, whose vertex is C, the triangle CDE will be the principal section of this cone (see Art. 114) (because its plane contains the line OP, which cuts the circular base perpendicularly at its centre), and as the plane of projection is evidently perpendicular to this principal section, and the angle CE'D' = CC'E = CDE, the projection is a subcontrary section of the cone, and therefore a circle by Art. 114.

If the circle to be projected passes through C, the projection is of course a right line.

2°. *If from any point on the sphere two tangents be drawn, the angle contained by them is projected into an equal angle.*

Let D be the point from which the tangents are drawn (see last figure). Now the tangent plane at D being perpendicular to OD, and the plane of projection being perpendicular to OC, the intersection of these planes is perpendicular to the plane of the triangle DOC, and therefore perpendicular to the line CD. Also, the angles OCD and ODC being equal, their complements are equal, that is, the angles made by the line CD with the tangent plane and the plane of projection are equal, and in opposite directions. It follows, then, by Art. 115, that the angle made by the tangents at D is equal to the projected angle.

3°. *If a cone be circumscribed to a sphere, the projection of its vertex is the centre of the circle which forms the stereographic projection of the circular curve of contact.*

In order to prove this Proposition, it is only necessary to show that the line D'E', in the preceding figure, is bisected by the right line CP' joining C, the centre of projection, to P', the point of intersection of two tangents at the points D and E. Now, since C is the pole of the tangent CK, with respect to the circle CPC', and P' the pole of DE with respect to the same circle, the line CP' is the polar of K, and therefore DK is cut harmonically, and consequently CK, CE, CP', CD make an harmonic pencil. It follows that D'E' (being parallel to CK) is (Art. 7) bisected, and the Proposition is established.

170. The first and third Propositions of the last Article may be deduced from Proposition 2° in the following manner:—

Any side of a right cone is evidently perpendicular to the corresponding tangent to its circular base, and therefore their projections, on the conditions of Art. 115, are at right angles. It follows at once from this consideration that the stereographic projection of the circular base of a cone circumscribed to a sphere is a curve of such a nature that all its tangents are at right angles to the right lines drawn to the points of contact from a certain point in its plane. That point is the projection of the vertex of the cone, and the right lines are the projections of its sides. Now it is evident that the circle is the only curve possessing the property just mentioned; and therefore the projection of a circle of the sphere, which may always be regarded as the base of a cone circumscribed, is another circle, having for its centre the projection of the vertex. Q. E. D.

When the circle to be projected passes through the centre of projection, the vertex of the cone lies in the plane touching the sphere at that point, and is therefore projected to infinity. *The projection is*, therefore, in this case, a circle having an infinite radius, that is, *a right line*.

INVERSE SURFACES.

171. The following Proposition is connected with our present subject :—

" If right lines be drawn from a given point to all the points of a given plane, and portions be taken on them, or on their productions through the point, which, measured from the point, are inversely proportional to the lines themselves; the extremities of these portions will, in either case, lie on the surface of a sphere passing through the given point, and having its centre on the perpendicular drawn from the point to the given plane."

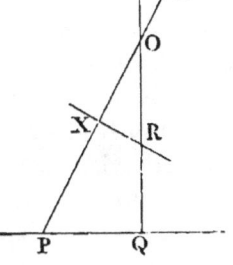

Let O (see fig.) be the given point, and P any point in the given plane, and let OP · OX = a constant quantity. Draw OQ perpendicular from O on the given plane, and on OQ take a portion OR, such that OR · OQ = OP · OX, and join R, X and

P, Q. The point R is fixed; and the (quadrilateral PQRX being evidently inscribable in a circle) the angle OXR = OQP, and is therefore a right angle. The locus of the point X is consequently (Euclid, B. iii. Prop. 31) a sphere described on OR as diameter.

If the portion OX' (see fig.) be measured from O in the opposite direction, OR' should be taken in a corresponding manner, and the locus will be a sphere with OR' as diameter.

If the locus of the point P were any other given surface or curve, there would be a corresponding surface or curve for the locus of X. A locus generated in this way from a given curve or surface has been named its *inverse*. The theorem above given may therefore be expressed as follows:—"The surface *inverse* to a plane is a sphere."

It also appears, from what has been proved, that the surface inverse to a sphere, when the given point O (which may be called the origin) is on the sphere, is a plane.*

172. Given in magnitude and position a circle; the locus of the centres of all the spheres, of which this circle may be regarded as a section, is evidently a right line drawn perpendicular to the plane of the circle at its centre. It follows from this remark that if, in addition to the circle, a point (not in the plane of the circle) be given, through which the sphere is to pass, the sphere is completely determined. We find, both in magnitude and position, a great circle of the sphere, by drawing through the given point and the locus of the centres a plane, and describing a circle through the given point and the two points in which this plane cuts the given circle.

173. *A sphere described through the circumference of the circular base of an oblique cone, and any point on its surface, cuts the cone in a subcontrary circular section* (see Art. 114).

In order to prove this Proposition we shall first lay down the following Lemma:—

LEMMA 18.—If from a given point O any right line be drawn cutting a given sphere in the points P and X, the rectangle OP · OX is constant.

---

* It is easy to infer from Art. 65, Prop. 30, that the surface inverse to a sphere is, *in general*, another sphere.

For (Lardner's Euclid, B. ii. Prop. 6) the rectangle is always equal to the difference of the squares of the radius and the line joining O to the centre.

Returning now to the original Proposition, it follows, from Art. 171, that the curve of intersection of the sphere and cone must lie on the sphere, which is *inverse* to the plane of the base of the cone, the vertex of the cone being the fixed point or *origin*. The curve in question is therefore *the intersection of two spheres, that is, a circle*. (The last assertion is readily proved as follows:—The chord of intersection of two circles is evidently perpendicular to the line joining the centres, and bisected by it; and if we conceive the circles to revolve on this latter line, they generate two spheres, whose curve of intersection is the circle described by the extremities of the chord.) Again, the section is subcontrary. For, in the first place, its plane is perpendicular to the principal section of the cone; since the centres of the two spheres above mentioned lie in the plane of the principal section. Secondly (see figure of Art. 114), if we suppose PQ to be the intersection of the principal section with the plane of the curve under consideration, the quadrilateral ABQP is inscribed in a circle, and therefore the angle PQC = the angle CAB; which proves the section to be subcontrary.

It follows immediately from what has been said, that a sphere can be made to pass through any two subcontrary sections of the cone.

It may be observed that the present Article furnishes an independent proof of the result given in Art. 114.

174. Two planes drawn through the vertex of the cone parallel to the plane of the circular base, and that of a subcontrary section, are called by Chasles the *cyclic planes* of the cone. This definition being laid down, the following Proposition is easily proved:—

A sphere passing through the circumference of the circular base of a cone, and through its vertex, touches the cyclic plane, which is parallel to the plane of the subcontrary section.

For this sphere may be considered (Art. 173) as cutting the cone in an infinitely small subcontrary circle; but since all parallel sections are similar, the cyclic plane in question is to be regarded as having exactly the same property; that cyclic plane has, therefore, an infinitely small circle in common with the sphere, that is, it touches the sphere.

And, conversely, it readily follows that *any sphere passing through the vertex; and touching one of the cyclic planes, cuts the cone in a circle whose plane is parallel to the other cyclic plane.*

An independent proof of Proposition 1° of Art. 169 follows at once from the first theorem given in this Article.

CHAPTER XII.

PROPERTIES OF PLANE AND SPHERICAL SECTIONS OF A CONE.

175. THE applications made of the method of projection in the preceding pages are, it is hoped, sufficient to enable the student to form a clear idea of its nature. The limits within which we have hitherto confined ourselves are, however, by no means adapted to exhibit the more general results to which the method is capable of leading. It seems proper, therefore, before bringing this work to a conclusion, to enter into some further development of this subject. In doing so, it will be necessary to presuppose, on the part of the reader (as we have done in the last Chapter), an acquaintance with a few of the elementary properties of the curves represented by the equation of the second degree in the method of co-ordinates. These curves are commonly called *conics*, as being identical with the sections of a cone made by a plane. It is, in fact, easy to prove, in the first place, that *every plane section of cone with a circular base* (in which sense it is to be recollected we have agreed to use the term) *is a curve of the second degree*, and, secondly, that *any given curve of the second degree may be regarded as a section of such a cone.*

In order to prove the first part of this Proposition, it is sufficient to observe that a circular section of the cone is intersected by a right line drawn at random in its plane in two points (which may perhaps become coincident or imaginary); for, any plane section of the cone being considered as the projection of this circle, possesses, in consequence, the same property; and this is the characteristic property of curves of the second degree.

(A more detailed proof, applied to the Elliptic, Hyberbolic, and Parabolic Sections separately, may be seen in Salmon's Conic Sections, Arts. 346 and 348.)

The second part of the Proposition above stated is included in the following theorem:—

176. *Any given curve of the second degree can be projected into a circle, whose centre is the projection of one of the foci.*

Let us take the ellipse for instance, and let (see fig.) AB be its major axis, CD its minor semi-axis, and F one of the foci. Through the major axis draw a plane perpendicular to the plane of the ellipse, and in this plane draw (in any direction) from either A or B (suppose B) a line *equal to the adjacent segment of the axis* made by F; let BF′ be this line, and produce it until F′A′ = BF′; join A′A and F′F, and let the joining lines meet in O. On A′B, as diameter, describe 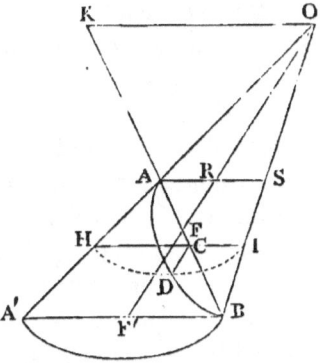 a circle in a plane perpendicular to the plane last mentioned, and make it the base of a cone with O as vertex; then, the given ellipse is a section of this cone.

For, through C draw a plane parallel to the base of the cone; this plane will cut the cone in a circle, having HI for its diameter, and will evidently pass through the axis minor of the ellipse. Now, if we draw AS parallel to A′B, we shall have CI = $\frac{1}{2}$AS = AR = AF (since BF′ = BF), and HC = $\frac{1}{2}$A′B = BF′ = BF, and therefore HC · CI = BF · FA = the square of the semi-axis minor CD (since F is one of the foci). It follows, that the point D lies on the circumference of the circle above mentioned, having HI for its diameter, and consequently lies on the surface of the cone. Again, it is evident that the section of the cone made by the plane of the given ellipse is an ellipse having for one of its axis AB, and the other coincident with CD. It appears then, finally, that this section and the given ellipse have their semi-axes equal and coincident, and therefore are identical. This proves the theorem in the case of the ellipse.

Should the curve to be projected be the parabola or hyperbola the same method still applies; although the demonstration above given will require to be modified, since, in the case of the parabola, the point D is at an infinite distance, and in that of the hyperbola is imaginary. The principle of continuity, however, teaches us (see Arts. 96 and 99) that these circumstances do not affect the correctness of the primary construction, and we shall therefore dispense with any further proof.

It is easy to prove that *in all three cases a plane drawn through the vertex of the cone and the directrix of the conic corresponding to the focus* F, *is parallel to the base of the cone.* For, if K be the point where the directrix meets the major axis, K and F (see the last figure) are harmonic conjugates with respect to A and B, and therefore OK must meet BA' in a point, the harmonic conjugate of F' with respect to A' and B, that is, at infinity. This proves the Proposition.*

Before concluding this Article, we may observe that by whatever method a *hyperbola* is projected into a circle, a plane drawn through the vertex of the cone, parallel to that of the hyperbola, divides the circle into two arcs, which are respectively the projections of the two distinct parts, of which the entire hyperbola is made up.

THE SPHERICAL ELLIPSE.

177. We are now in a position to transfer immediately many of the most general properties of the circle to all the curves of the second degree; but as there are other curves (of which the latter are only a particular case), to which the properties of the circle, above referred to, admit of being transferred with almost equal facility, it seems proper, before proceeding further, to explain the

---

* Poncelet has given a method by which a curve of the second degree may be projected into a circle, so that *any given point* in the plane of the curve shall be projected into the centre of the circle. The particular mode of projection adopted in the text possesses, however, certain peculiar properties which will be useful in the sequel (see Art. 188).

nature of those curves, at least so far as shall be required for the due application of the elementary principles already laid down.

The curves in question are those formed by the intersection of a cone with a sphere, whose centre coincides with the vertex. From the mode of their formation they are called *spherical conics*; and they are, in general, of double curvature, that is, not plane. In order to be plane curves, they must, of course, be circles, which only happens when the cone is right.

By producing the sides of the cone through the centre of the sphere, we have two distinct parts of the complete curve of intersection; but it will be sufficient at present to confine our attention to one of the portions, each of which is usually called a *spherical ellipse*.

In explaining the nature of this curve we shall not assume any property of the cone besides those established in the preceding pages. This remark will account for some peculiarities in our mode of treating the subject.

Our fundamental theorem is derived from the following Proposition relating to the cyclic planes of the cone:—

The plane containing any two sides of a cone intersects the cyclic planes in two right lines, which make equal angles with the two sides respectively. (Chasles's Memoir on Cones, Art. 22.)

Let OA and OB (see fig.) be the two sides of the cone, and ACO the circle in which their plane cuts the sphere, which passes through the vertex O, and the circumference of the base, and (Art. 174) touches the subcontrary cyclic plane. Let OC′ be the line in which the plane of this circle cuts the tangent plane; OC′ is therefore a tangent; also, if OC be the line in which the circle cuts the other cyclic plane, OC is parallel to AB. We have then the angle C′OB = OAB = COA. Q. E. D.

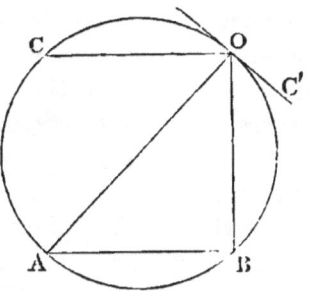

178. The theorem above referred to is as follows:—

If a great circle be drawn cutting a spherical ellipse, the portions of it between the points of intersection and the two cyclic arcs respectively, are equal. (The *cyclic arcs* of a spherical conic are the two great circles whose planes are the cyclic planes of the cone.) (Chasles's Memoir in Spherical Conics, Art. 13.)

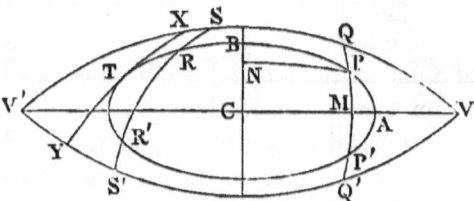

The theorem follows at once from the Proposition proved in the last Article, since arcs on the sphere are the measures of angles subtended by them at the centre, which is here the vertex of the cone.

The figure is intended to represent a spherical ellipse, with its two cyclic arcs, SQ, S'Q'. SS' represents any great circle cutting the ellipse and its cyclic arcs, and making the segments SR and S'R' equal to one another, according to the theorem just given.

When the points R and R' coincide, the great circle touches the curve, and we have the following result:—*If a great circle touch a spherical ellipse, the portion of it intercepted by the cyclic arcs is bisected by the point of contact.*

179. The results arrived at in the last Article lead to important consequences.

1°. If a great circle, CA (see last figure), be drawn, *bisecting the angle made by the cyclic arcs*, and another, QQ', be drawn at right angles to this, and cutting it at M, we have evidently QM = Q'M; but also (Art. 178) QP = Q'P'; and therefore PM = P'M. It follows from this that *the curve is symmetrical with respect to the great circle* CA.

Again, *if* BC *represent a great circle at right angles to the two cyclic arcs the curve is also symmetrical with respect to it,* since the cone is evidently symmetrical with respect to the plane of this great circle (which is, in fact, the plane of the principal section of the cone.) (See Art. 114.)

The portions CA and CB of the great circles above described may be called the *semi-axes* of the spherical ellipse, and expressed by the letters a and b respectively. The figure represents CA as greater than CB; this is proved to be the case in Art. 181, 1°.

The point C is evidently the middle point of the semicircular arc VV'. It is also easy to see, from what has been said, that if any great circle be drawn through C, the part intercepted by the curve is bisected at C; this point may therefore be called the *centre* of the spherical ellipse.

2°. *The projection of the curve on a plane touching the sphere at C is evidently an ellipse* (Art. 175). It follows from 1° that its centre is C, and that its semi-axes are tan a and tan b (unity being the radius of the sphere), the projections of the arcs CA and CB.

These considerations enable us *to find* at once *the equation of the curve, whether in rectangular or polar co-ordinates*. To find the former, let (see last figure) PM and PN be arcs drawn from any point P of the curve at right angles to its axis (which we shall take as axes of co-ordinates), and let the arcs CM and CN be called x and y respectively. Now, since the equation of a plane ellipse, referred to its axes, is of the form

$$\frac{x^2}{a^2} + \frac{y^2}{b^2} = 1,$$

it follows from Art. 109 that the required equation is

$$\frac{\tan^2 x}{\tan^2 a} + \frac{\tan^2 y}{\tan^2 b} = 1.$$

Again, if we conceive the points C and P to be joined by an arc, which we may call ρ, and if we represent by θ the angle made by this arc with the semi-axis CA, we shall have the following polar equation of the spherical ellipse:—

$$\frac{\cos^2 \theta}{\tan^2 a} + \frac{\sin^2 \theta}{\tan^2 b} = \frac{1}{\tan^2 \rho} = \cot^2 \rho.$$

This comes at once from the corresponding equation of the plane ellipse, which is of the form

$$\frac{\cos^2 \theta}{a^2} + \frac{\sin^2 \theta}{b^2} = \frac{1}{\rho^2}.$$

3°. We saw in the last Article that a tangent arc, such as XY (see last figure), intercepted by the cyclic arcs, is bisected at the point of contact T. From this is derived one of the most remarkable properties of the curves under consideration.

The area of the spherical triangle contained by the cyclic arcs and any tangent arc is constant.

The proof which we shall give of this Proposition is taken from Professor Graves's *additions* contained in his translation of Chasles's Memoirs on Cones and Spherical Conics.

Let XY and X'Y' (see fig.) be two infinitely close positions of the tangent arc; let O be their point of intersection, and T and T' the points of contact. Now, since TO and T'O are infinitely small, we have OX = OY, and OX' = OY', and therefore the triangles OXX' and OYY' are equal, and consequently the triangle V'XY = V'X'Y'. This proves that the infinitesimal change on the triangle V'XY = 0, and therefore that its area remains constant when the point T gradually changes its position along the curve.

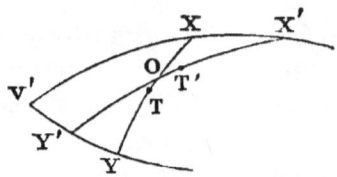

Hence we infer conversely, that *the envelope of the base of a spherical triangle, whose vertical angle and area are given, is a spherical ellipse, whose cyclic arcs are the two arcs forming the given vertical angle.*

Since the area of a spherical triangle depends on the sum of its three angles, we have also the following important theorem:—

The sum of the angles made, by any arc touching a spherical ellipse, with the cyclic arcs is constant.

In the preceding figure the angles to be taken are V'XY and V'YX. If the supplement of one of these angles be taken in place of the angle itself, the *sum* must be changed into the *difference* in the enunciation of the theorem.

180. Before proceeding further we shall require the following Lemma:—

LEMMA 19.—*The reciprocal of an ellipse, with respect to a concentric circle* (see Art. 42, 4°), *is another ellipse, whose axes are coincident with those of the former, and inversely proportioned to them.*

Let CP (see figure on next page) be the perpendicular from the centre C of the given ellipse on any tangent, and let CP · CO = k^2, k being the radius of the circle (which is not drawn). If CD be drawn parallel to the tangent, we have, by a well-known property of the ellipse, $CP \cdot CD$ = the rectangle under the semi-axes of the ellipse = a constant quantity, and therefore $\dfrac{CO}{CD}$ = constant. It 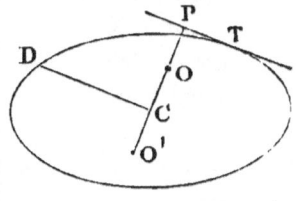 follows at once that the locus of O (the reciprocal curve) is an ellipse similar to the given one, but not similarly situated, the corresponding radii vectores CO and CD being at right angles. This proves the Proposition above enunciated.

It is to be observed, that this result is not affected by the sign of k^2. If k^2 be negative, we must take CO' (see fig.) $= CO$; but the locus will be the same as before.

We shall now give the corresponding theorem on the sphere.

The reciprocal of a spherical ellipse (see Art. 143) *is another curve of the same kind, whose semi-axes are coincident with of the former, and respectively their complements.*

This appears at once from the foregoing Lemma, combined with Art. 145, 1°, and Art. 179, 2°.

181. It appears from Art 179, 3°, that a spherical ellipse may be regarded as the envelope of the base of a spherical triangle, in which the vertical angle is given, and also the sum of the base angles. By reciprocating this theorem (see Art. 148), we find (Art. 180) that *a spherical ellipse may be regarded as the locus of the vertex of a spherical triangle, whose base is given, and also the sum of the sides.* In other words, when a spherical ellipse is given, two fixed points exist on the surface of the sphere, the sum of the arcs from which, to any point on the curve, is constant. These points are called its *foci*, and possess remarkable analogies to the foci of a plane ellipse. We shall now proceed to the consideration of a few of their properties.

1°. From the mode by which we have ascertained the existence of these points, it appears that they are *the trigonometrical poles of the cyclic arcs of the supplementary ellipse.* Hence Art. 180), *the foci of a spherical ellipse lie on one of its axes at equal distances from the centre;* and it is easy to prove that *that axis must be the greater.*

2 A

Let F and F′ (see fig.) be the foci, and P any point on the curve, we have FP + F′P = constant, and FC = F′C, C being the centre. (Since the base of a spherical triangle is less than the sum of the sides, it is evident that F and F′ lie within the curve, as the figure represents.)

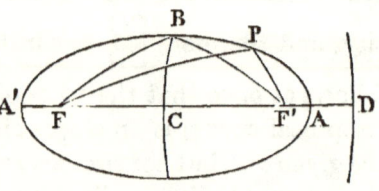

By supposing P to coincide with the ends of the semi-axes, A and B, successively, we have, for the value of the constant, FA + F′A, or FB + F′B, that is, 2CA or 2FB; and therefore CA = FB. Now cos FB = cos FC · cos CB, and therefore cos FB is less than cos CB, that is, cos CA is less than cos CB, from which it follows that CA is greater than CB. Q. E. D.

Since (Art. 180) the coincident semi-axes of two supplementary spherical ellipses are complements, it appears, from what has just been proved, that *the minor axis of a spherical ellipse is that which is perpendicular to its cyclic arcs* (see the figure of Art. 178).

When the semi-axes a and b are given, the position of the foci is determined by the equation $\cos c = \dfrac{\cos a}{\cos b}$, where c represents the distance FC. The position of the cyclic arcs is also readily determined. For if we call ϕ the semi-angle made by them, and a', b', c' the quantities in the supplementary ellipse, analogous to a, b, c, we shall have the following equations:—

$$2\phi + 2c' = \pi, \quad a' + b = \tfrac{\pi}{2}, \quad b' + a = \tfrac{\pi}{2}, \quad \text{and } \cos c' = \dfrac{\cos a'}{\cos b'},$$

and therefore
$$\sin \phi = \dfrac{\sin b}{\sin a}.$$

The geometrical meaning of this result is, that *the angle made by the cyclic arcs is equal to that which the minor axis subtends at either of the foci.*

2°. If we reciprocate the theorem given in Art. 178, we find that *the two arcs drawn from the foci to the point of intersection of two tangent arcs make equal angles with these tangent arcs respectively.*

The angles FPT and F′PT′ (see figure on next page) are the angles to be taken.

If the points T and T' coincide, we find that *a tangent arc at any point, suppose T, bisects the external vertical angle of the triangle, whose vertex is the point of contact, and base the arc joining the foci.*

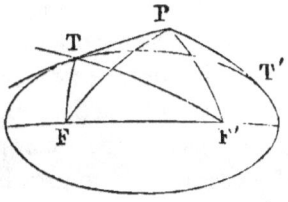

3°. It follows, from Art. 178, that if two spherical ellipses have the same semi-circular cyclic arcs, any arc touching the inner curve meets the outer in two points, which are equally distant from the point of contact. For, since TX = TY (see fig.) and NX = OY, TN = TO.

It can be proved from this, by reasoning similar to that contained in Art. 179, 3° that *the spherical area of the segment NSO (see fig.) cut off from the outer curve, is constant* when the point of contact, T, changes its position.

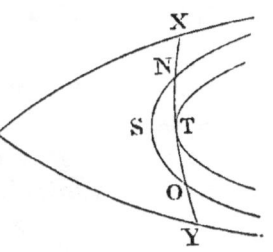

If we reciprocate this result, we find (Art. 147) that *if two spherical ellipses have the same foci, and if from any point in the outer two tangent arcs be drawn to the inner curve, the sum of these arcs, and of the concave part of the circumference of the curve included between them, is constant.*

This theorem is due to Professer Graves. (See his translation of Chasles's Memoirs on Cones and Spherical Conics, p. 77.)

4°. If we draw any side of a cone, and drop perpendiculars from the vertex on the planes of the base, and of a fixed subcontrary section, the product of the sines of the angles made by the side with the two planes is equal to the product of the perpendiculars divided by the product of the portions of the side between the vertex and the two planes; but the latter product is constant (Art. 173, Lemma 18), since the two sections made by the planes lie on a sphere (Art. 173). We have then the following theorem:—"The product of the sines of the angles made by any side of a given cone with the cyclic planes is constant."

From this we infer that *if two arcs be drawn from any point on a spherical ellipse, at right angles to the cyclic arcs, the product of their sines is constant.*

If this be reciprocated, we find that *the product of the sines of arcs drawn from the foci perpendicular to any tangent arc is constant.*

182. The theorem, whose reciprocal has just been given, may be put in another form, which leads to results of some importance. The sines of the perpendiculars on the cyclic arcs being equal to the cosines of their complements, we may say that the product of the cosines of the arcs between any point on the curve and the trigonometrical poles of the cyclic arcs is constant.

Hence it appears that *if the base of a spherical triangle be be given, and the product of the cosines of the sides, the locus of the vertex is a spherical ellipse, whose cyclic arcs are the great circles whose poles are the extremities of the base.*

Since the cosine of the hypotenuse of a right-angled spherical triangle is equal to the product of the cosines of the sides, the locus of the vertex of such a triangle, whose hypotenuse is given, is a spherical ellipse.

From this again it follows, that *the locus of the vertex of a spherical triangle, in which the base is given, and the ratio of the sines of the sides, is a spherical ellipse* (Art. 139, Lemma 14).

183. We shall place here a Proposition which is of use in the geometrical investigation of the properties of Fresnel's *wave surface.*

"If the vertex of a right angle be fixed, and one leg move on a given plane, while the plane of the right angle turns on a given right line, the other leg describes a cone, one of whose cyclic planes is the given plane, and the other a plane perpendicular to the given line."

If we describe a sphere having the vertex of the right angle as centre, and any radius (taken for unity), the theorem to be proved evidently comes to the following:—

If a quadrantal arc on the sphere passes through a fixed point, while one of its extremities moves on a given great circle, the locus of the other extremity is a spherical ellipse, one of whose cyclic arcs is the given great circle, and the other a great circle whose pole is the given point.

Let A be the given point, and OX the quadrantal arc, one of whose extremities X moves on the given great circle XB (see fig.); let AB be drawn perpendicular to the given great circle, and be produced to P, its pole; and let P, X, and P, O, be joined by arcs. Then, since PX and OX are quadrants, X is the pole of the arc PO, and POA is a right-angled triangle, whose hypotenuse AP is given completely; the locus of O is therefore (Art. 182) a spherical ellipse, whose cyclic arcs have for poles the points P and A; which proves the Proposition.

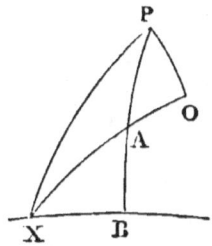

THE SPHERICAL HYPERBOLA.

184. We have hitherto been considering only one of the distinct portions, which together constitute the spherical conic; there is, however, another mode of dividing the curve symmetrically, which must not be omitted. This is by means of the great circle at right angles to the cyclic arcs.

In the annexed figure, BC and B'C' represent two parts of the great circle referred to; and the two semi-ellipses lying between

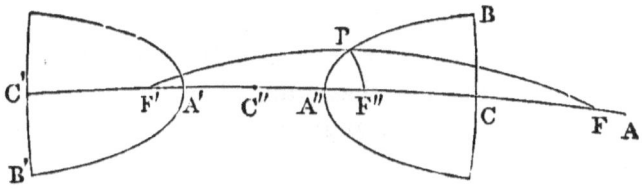

C and C' are on one of the hemispheres formed by it. These two, taken together, are called by Chasles a *spherical hyperbola*. Its foci are F" and F' a point diametrically opposite to F, (F and F" being the foci of one of the spherical ellipses), and possess the following fundamental property:—

The difference of the arcs joining the foci of a spherical hyperbola to any point on the curve is constant.

Let P be the point. Now since F' is diametrically opposite to F, F'P and FP are portions of the same great circle, and therefore F'P + FP = a semi-circle; but FP + F"P = constant (Art. 181); therefore, by subtraction, F'P − F"P = constant. *Q. E. D.*

The value of the constant is equal to the supplement of the arc $A A'' = A'A'' =$ twice $C''A''$. (The point C'' bisects $A'A''$, and may be called the centre of the spherical hyperbola).

The analogy subsisting between the cyclic arcs of the spherical conic and the asymptotes of a plane hyperbola cannot have escaped the notice of the reader.

185. We shall conclude this part of our subject with the following important theorem:—

If two spherical conics have the same foci, and cut one another, they intersect at right angles.

It is easy to see, in the first place, that two spherical conics having the same foci cannot intersect unless they are so situated on the sphere, that two foci belonging to the one, when considered as a spherical ellipse, are the foci belonging to the other, considered as a spherical hyperbola; for if they were both ellipses or both hyperbolas, they could not have one point in common without coinciding throughout. This being understood, let P be one of their points of intersection, and let arcs be drawn from P to the two foci, F and F'. Now if an arc be drawn touching the spherical ellipse at P, it will (Art. 181, 2°) bisect the external vertical angle of the triangle FPF', and an arc touching the hyperbola at the same point evidently bisects the vertical angle FPF' itself (Art. 184); therefore these tangent arcs are at right angles to one another, that is, the angle made by the two curves is a right angle.

PROJECTIVE PROPERTIES OF PLANE AND SPHERICAL CONICS.

186. We shall now state some of the general properties of plane and spherical conics, deducible from those of the circle by the method of projection.

From the definitions of these curves, it appears that *they may always be considered as projections of the circular sections of a cone;* hence we have the following results:—

1°. All merely *graphical* properties of the circle (see Art. 113) are true for plane or spherical conics; great circles being substituted in the latter case for right lines. Pascal's theorem (see Art. 27) and Brianchon's theorem (see Art. 42, 1°) may be taken as examples.

2°. Properties of the circle, of the kind stated in Art. 110, are at once transferred to any plane or spherical conic. The following theorem, due to Carnot, will serve as an example:—

"If all the sides of a plane polygon be cut by a curve of the second degree, the continued product of the distances from each angle to the points of intersection of the consecutive side with the curve is equal to the analogous product formed by inverting the successive order of the sides and angles."

Now, this equation evidently holds good in the case of a circle (Euclid, B. iii. Props. 35 and 36); it is therefore true (Art. 110) for any conic section; that is, for any curve of the second degree (Art. 176).

In order to adapt the Proposition to the case of a spherical conic, we have (Art. 110) only to substitute a spherical polygon in place of the plane, and the *sines* of the distances in place of the distances themselves.

If the sides of the polygon become tangents, we have, *in plano, the product of one set of alternate segments equal to that of the other set*, and a corresponding theorem on the sphere.

3°. We saw, in Art. 128, that the anharmonic ratio of four planes passing through four fixed sides of a cone, and intersecting in a fifth variable side, is constant: hence (Art. 116), *in any curve of the second degree, the anharmonic ratio of four right lines drawn from four fixed points on the curve to a variable fifth point is constant;* and (Art. 118), *the anharmonic ratio of the spherical pencil formed by arcs drawn from any point of a spherical conic to four fixed points on the curve is also constant.* (This constant ratio is, in both cases, called the *anharmonic ratio of the four points.*)

4°. It was proved in Art. 42, 3°, that the anharmonic ratio of the four points, in which a variable fifth tangent to a circle is cut by four fixed tangents, is constant: hence follows immediately (Art. 110) a corresponding property of any plane or spherical conic. In the latter case, of course, the anharmonic ratio of four points on a great circle (see Art. 120) must be taken in the enunciation.

5°. The Proposition proved in Art. 39 may be otherwise expressed:—

"If through a given point a right line be drawn, the locus

of the harmonic conjugate of this point with respect to the two points in which the line cuts a given circle, is a right line."

From this it follows (Art. 7), that a similar enunciation holds good for any curve of the second degree; and the given point may be called a *pole*, and the locus of the harmonic conjugates its *polar* in relation to the curve.

Again, it is equally evident (Art. 121) that the locus of the harmonic conjugates of a given point on the sphere, with respect to the two points in which an arc through it cuts a spherical conic, is a great circle, which also may be called the *polar of the point* in relation to the conic.

These definitions being laid down, we may assert that *a point and its polar, with respect to any plane or spherical conic, may be regarded as the projections of a point and its polar, in relation to a circle;* and by this consideration most of the fundamental Propositions in the theory of polars, already established for that particular case, are extended to the more general curves just mentioned. It is important to observe, in particular, that all the theorems contained in Art. 40 are included in this remark; great circles, being, of course, substituted in place of right lines, in the case of the spherical conic.

It is hardly requisite to remark, that a tangent to a plane conic is the polar of the point of contact, and the chord of contact of two tangents the polar of their intersection. Corresponding properties obtain, of course, in the case of the spherical conic.

Since any right line or arc through the *centre* of a plane or spherical conic is bisected, it follows, from the definitions above given, that *the polar of the centre is, in the former case, a right line at infinity, and, in the latter, a great circle having that point for its trigonometrical pole.* And conversely, a point whose polar satisfies either of these conditions is the centre of the plane or spherical conic.

In place of considering a plane or spherical conic as the projection of a circle, it is sometimes convenient *to regard one as the projection of the other.* In such cases, it is evident that *a point and its polar are still projected into a point and its polar.*

187. In the case of the circle, the right line joining the centre

to the pole O (see Art. 38) is perpendicular to the polar, and the rectangle $CO \cdot CO' =$ the square of the radius; let us examine to what extent similar properties hold good for the more general curves.

1°. *When the conic is plane:* let C be its centre, O any point, and O'M its polar; then, since O'B' (see fig.) is cut harmonically (Art. 186, 5°), we have (Art. 3) $CO \cdot CO' = CA'^2$. This equation expresses that *the semi-diameter on which the pole lies is a mean proportional between the distances along the semi-diameter from the centre to the pole and polar.*

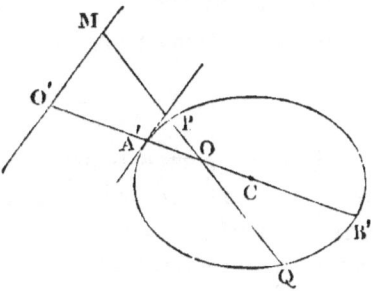

In order to determine the inclination of the polar to the diameter CO, let us conceive the chord PQ to revolve round O, until it becomes parallel to the polar, and let P'Q' be the chord in that position; P'Q' is bisected (Art. 3), since the harmonic conjugate of O, with respect to P' and Q', is at infinity; it is therefore one of the chords belonging to the diameter CO, and therefore parallel to the tangent at A'. This proves that *the polar of a point is parallel to the tangent at the extremity of the diameter passing through the point.* It follows that *the polar is at right angles to the line CO, only when the point O lies on one of the axes of the conic.*

2°. *When the conic is spherical:* let us imagine the spherical figure, corresponding to that above given, to be projected on the plane touching the sphere at C; we shall then have (Art. 179, 2°) a plane conic, whose centre is C, and whose semi-axis are the projections of those of the spherical curve. The point O and its polar, in relation to the latter curve, will (Art. 186, 5°) be projected into a point, and its polar with respect to the former; and hence we infer, first, that $\tan CO \cdot \tan CO' = \tan^2 CA'$; in the next place, that *the polar O'M intersects the arc touching the curve at A' in a point on the sphere, whose distance from C is a quadrant* (because its tangent is infinite); and finally (Art 109), that *the polar O'M is perpendicular to the arc CO, only when the point O lies on one of the axes of the curve.*

2 B

When the point O coincides with one of the foci of the spherical conic, its polar in relation to the curve is called by Chasles the *director arc* corresponding to that focus, and possesses properties analogous to those of the directrix of a plane conic. The director arc corresponding to the focus F' (see figure of Article 181, 1°) is perpendicular to the major axis at a point D, whose distance from the centre is determined by the equation

$$\tan d = \frac{\tan^2 a}{\tan c},$$ where d represents the arc CD.

188. Properties relating to angles at a focus of a plane conic may be deduced with great facility from those concerning angles at the centre of a circle, by the aid of the theorems established in Arts. 115 and 176. The construction employed in the latter Article is such (Art. 115) that *an angle, whose vertex is at the focus of the conic, is projected into an equal angle, having its vertex at the centre of the circle.* The following are examples:—

1°. Two tangents drawn from a point to a circle subtend equal angles at the centre. Hence we infer that two tangents drawn from a point to any curve of the second degree subtend equal angles at one of the foci. (This Proposition has been already given in Art. 168.)

2°. If a chord of a circle whose radius is r, be drawn through a point whose distance from the centre is d, it is easily proved that the product of the tangents of the semi-angles subtended at the centre by the segments of the chord $= \dfrac{r-d}{r+d}$ (see Lemma 16 of Art. 148, 4°). Hence it follows, that *the product of the tangents of the halves of the angles subtended at one of the foci of a given conic by the segments of a chord drawn through a given point is constant.*

3°. A right line drawn from the centre of a circle parallel to a chord bisects the external vertical angle of the isosceles triangle, whose sides are the radii passing through the extremities of the chord. From this we have the following theorem:—
"The right line joining one of the foci of a conic to the point where the corresponding directrix is cut by any chord of the

conic bisects the external angle formed by lines from the focus through the extremities of the chord."

4°. From what has been proved in Art. 176 it readily follows, that *concentric circles may be projected into conics having one focus and the corresponding directrix common.*

In the annexed figure (which corresponds to that in Art. 176), F'B and F'B" represent the radii of the concentric circles; and F"'B', which is parallel to F'B, is equal to FB', since (Art. 176) F'B = FB; the circle whose radius is F"'B', and whose plane is parallel to that of the former circles, is then projected into a conic, one of whose foci is F, and whose corresponding directrix passes through K. This proves the Proposition.

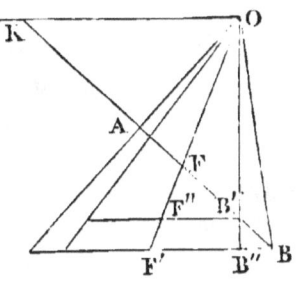

In order to illustrate the application of this principle, let us return to Example 2°, and let us inquire *how the point through which the chord of the conic is drawn must move, in order that the product of the tangents of the semi-angles subtended at the focus may retain the same constant value.* Since the value of the product in the case of the circle depends on the distance d, the answer evidently is, that the point must move on another conic, having one focus and the corresponding directrix the same as those of the given conic.

5°. A chord of a circle which subtends a constant angle at the centre touches a concentric circle; and tangents at its extremities intersect on a third circle, also concentric with the original. Hence we see that *if a chord of a given conic subtend a constant angle at one of the foci, its envelope is another conic, and the locus of the intersection of tangents at its extremities is a third conic; all three conics having, moreover, one focus, and the corresponding directrix in common.*

189. The following principle is of use in deriving properties of a focus in the case of spherical conics from the analogous properties of the plane curves.

The projection of a spherical conic on a plane touching the sphere at one of the foci is a plane conic, of which also the point of contact is a focus.

Since the foci of a spherical conic are the trigonometrical poles of the cyclic arcs of the supplementary conic (Art. 181, 1°), the projection of the latter on the tangent plane is a circle; but the projection of the original conic is the polar reciprocal (Art. 145, 1°) of the former projection with respect to a circle whose centre is the point of contact; it is therefore (Art. 167) a curve of the second degree, having one of its foci coincident with that point. Q. E. D.

It is evident that *the directrix of the plane conic, corresponding to the point of contact, is the projection of the corresponding director arc* (see Art. 187, 2°) of the spherical curve; because a point and its polar in relation to the latter are projected into a point, and its polar with respect to the former.

Since all parallel sections of a cone are similar, it follows from the property above proved, that the right lines drawn from the centre of the sphere to the foci of a spherical conic are such that any plane perpendicular to either is cut by it in a point, which is one of the foci of the corresponding section of the cone, by whose intersection with the sphere the spherical conic is formed. These right lines are called the *focal lines* of the cone, and are of very great importance in connexion with the general theory of surfaces of the second degree. *They are* (Art. 181, 1°) *perpendicular to the cyclic planes of the supplementary cone.*

190. The principle stated at the commencement of the last Article may also be derived from *the polar equation of a spherical conic referred to one of the foci* as pole.

In order to find the equation, let FP (see figure of Art. 181, 1°) be represented by ρ, and the angle PFF′ by θ; we have (Art. 181, 1°) $F'P = 2a - \rho$, and

$$\cos F'P = \cos FP \cdot \cos FF' + \sin FP \cdot \sin FF' \cos PFF';$$

and therefore

$$\cos 2a \cdot \cos \rho + \sin 2a \cdot \sin \rho = \cos \rho \cdot \cos 2c + \sin \rho \cdot \sin 2c \cdot \cos \theta,$$

or, $(\cos 2a - \cos 2c) \cos \rho = (\sin 2c \cdot \cos \theta - \sin 2a) \sin \rho,$

and finally,
$$\tan \rho = \frac{\cos 2c - \cos 2a}{\sin 2a - \sin 2c \cdot \cos \theta}.$$

Since $\tan \rho$ is the projection of ρ, and the angle θ remains unaltered (being the angle made by tangents at F), this result proves that the projection of the curve on a plane touching the sphere at F is a plane conic, having F for one of its foci, and having an excentricity equal to $\dfrac{\sin 2c}{\sin 2a}$.

191. We shall now give some examples of the process of deduction alluded to in Art. 189.

1°. If ρ_1 and ρ_2 be the segments of a focal chord of a given plane conic, it is known that $\dfrac{1}{\rho_1} + \dfrac{1}{\rho_2} = $ constant; what is the analogous property of a spherical conic?

The required Proposition is:—" The sum of the cotangents of the segments of a spherical chord through one of the foci is constant."

This result may, of course, be derived directly from the polar equation given in the last Article.

2°. *What is the property of a director arc* (see Art. 187, 2°) *of a spherical conic, corresponding to the property of plane conics, from which the directrix derives its name?*

Let O (see fig.) be the centre of the sphere, and P' the projection of P, any point of the curve on a plane touching the sphere at the focus F; let DV be the director arc corresponding to F, V being the foot of a perpendicular arc on it from P; let D'X be the projection of the director arc, X being the foot of a perpendicular drawn from P' on this projection; let δ represent the angle made by the tangent plane with that of the director arc, and p a perpendicular from P' on the latter plane. Then since (Art. 189) F is the focus, and D'X

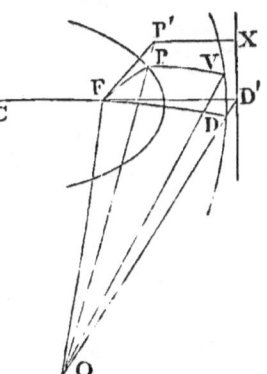

the corresponding directrix of the plane conic, which constitutes the projection of the given curve, $\dfrac{FP'}{P'X} = $ constant; now

FP' = OP' · sin FOP' (since OFP' is a right angle), and therefore

= OP' · sin FP (the radius of the sphere being unity),

and $\quad P'X = \dfrac{p}{\sin \delta} = \dfrac{OP' \cdot \sin POV}{\sin \delta} = \dfrac{OP' \cdot \sin PV}{\sin \delta}$;

we have therefore, by division,

$\dfrac{\sin FP \cdot \sin \delta}{\sin PV} =$ constant, and consequently $\dfrac{\sin FP}{\sin PV} =$ constant,

which is the required result.

3°. From Example 1° of Art. 188, we infer that *two arcs drawn from a point to touch a spherical conic subtend equal angles at one of the foci.*

4°. The second and fourth Examples of the same Article lead to the following results:—" If an arc be drawn through a given point, cutting a given spherical conic, the product of the tangents of the semi-angles, subtended by its segments at one of the foci, is constant."

" The product of the tangents will retain the same constant value when the point moves on another spherical conic having one focus and the corresponding director arc in common with the given conic."

5°. In a similar manner we find, that "the arc joining one of the foci of a spherical conic to the point where the corresponding director arc is cut by any spherical chord bisects the external angle formed by arcs drawn from the focus to the extremities of the chord."

6°. "If a spherical chord of a spherical conic subtend a constant angle at one of the foci, its envelope is another spherical conic; and the locus of the intersection of tangent arcs drawn at the extremities of the chord is also a spherical conic; and all three curves have one focus, and the corresponding director arc in common." This appears from Art. 188, 5°.

192. In obtaining properties of spherical conics by projection from those of the circle, or any other plane conic, it is important to observe, that the polar of the centre of the plane curve is a right line at infinity (Art. 186, 5°), and may therefore be considered as the line of intersection of the plane of the curve, with a parallel plane drawn through the centre of the sphere. It follows from this remark (Art. 186, 5°), that the projection of the centre of the given plane curve is the pole, in relation to the spherical conic, of the great circle formed by the parallel plane.

It is also of importance to remark, that if two planes be drawn through the centre of the sphere and any two right lines in the given plane figure, these planes will intersect the parallel plane above mentioned in two radii containing an angle equal to that made by the two right lines.

When the spherical conic is (as we shall generally suppose) considered as the projection of a circle, the parallel plane through the centre of the sphere becomes a cyclic plane of the cone, *and the centre of the circle is projected into the pole of the corresponding cyclic arc in relation to the spherical conic.*

We shall now illustrate the application of these remarks by some examples, referring the reader for further details to Graves's translation of Chasles's Memoirs on Cones and Spherical Conics, before cited.

1°. A radius of a circle is perpendicular to a tangent at its extremity. Hence we find that an arc touching a spherical conic, and the arc drawn through its point of contact and through the pole of a cyclic arc, with relation to the conic, meet that cyclic arc in two points, the distance between which is a quadrant.

2°. Two tangents to a circle make equal angles with the chord of contact. Hence we infer that two tangent arcs to a spherical conic, and the arc which joins their points of contact, intersect a cyclic arc in three points, the third of which bisects the distance between the first two.

3°. Angles in the same segment of a circle are equal. This leads to the following:—If through two fixed points on a spherical conic two arcs be drawn, which intersect in any third point of the curve, the segment which they intercept upon a cyclic arc is of invariable magnitude.

4°. The vertex of an angle of invariable magnitude, whose sides touch a given circle, generates a second circle; and the envelope of the chord of contact is a third circle; and these three circles are concentric. Hence we infer that, if two tangent arcs be drawn to a spherical conic, so that the segment intercepted between them upon a cyclic arc may be of a constant length, the locus of their point of concourse will be a second spherical conic. The arc joining the two points of contact will envelope a third conic. The cyclic arc in question will be a cyclic arc of the two new conics, and this arc will have the same pole with relation to the three conics.

5°. If right lines be drawn from two fixed points in the circumference of a circle through the extremities of any diameter, they will intersect in a point, the locus of which will be a circle which passes through the two fixed points, and whose centre is the pole of the right line joining them. The student will not find it difficult to prove this theorem. From this is derived the following:—If arcs be drawn from two fixed points on a spherical conic to the extremities of any spherical chord passing through the pole of a cyclic arc, with relation to the curve, they will intersect in a point, the locus of which will be a second spherical conic; the cyclic arc in question will be a cyclic arc of the new conic, and its pole with relation to that curve will be the same as the pole with relation to the given conic of the arc joining the two fixed points.

193. In the examples just given, we have employed a circular section and the parallel cyclic plane of the cone, by whose intersection with the sphere the spherical conic is formed. Let us now suppose the cone to be cut by any plane; this will in general give an ellipse or hyperbola; and it is known that the vertex of a right angle circumscribed to either of these curves lies upon a circle which has the same centre with it. Let us also draw, as before, a parallel plane through the centre; this will form a great circle, two of whose radii will be parallel to the legs of the right angle, and each pair of parallel lines will determine the plane of a great circle touching the spherical conic. Hence we obtain the following theorem:—" A spherical conic and a fixed arc arbitrarily drawn being given, if two tangent arcs to the conic be drawn, so that the segment intercepted between them on the given arc may be a quadrant, the point of concourse of these two arcs will generate a second conic, which will have the given arc for a cyclic arc. And this arc will have the same pole in relation to the two conics."

If the plane cutting the cone be parallel to a tangent plane, the plane section becomes a parabola, and the locus of the intersection of rectangular tangents to it a right line. This particular case leads to a corresponding result, as follows:—

"If two tangent arcs to a spherical conic intercept between

them, a segment equal to a quadrant, on a fixed tangent arc to the curve, the point of concourse of the two tangent arcs will move along an arc of a great circle." A particular case of this result is given in Art. 148, 4°.

194. In order to reciprocate properties, such as those arrived at in the last two Articles, we shall require the following principle:—

To a point and its polar arc, in relation to a spherical conic, there correspond an arc and its pole in relation to the supplementary conic. (A point and a great circle are here said to *correspond* when the point is the trignometrical pole of the great circle. See Art. 143.)

In order to prove this Proposition, let us first suppose that the polar arc in one conic joins the point of contact of two real tangent arcs drawn from the pole; these points of contact (Art. 143) correspond to two great circles touching the supplementary conic, and therefore the polar arc in question corresponds to the point of intersection of these two great circles; this proves (Art. 186, 5°) the Proposition in the case supposed.

When the polar arc in one conic does not meet the curve, the foregoing proof will not apply; but in virtue of the principle of continuity (Art. 99), we may extend the result to this case also.

The following is a particular case of the theorem above given:—

To a cyclic arc and its pole in relation to a spherical conic, there correspond a focus and its director arc in respect to the supplementary conic.

195. The reciprocals of the theorems given in Art. 192 are as follow:—

1°. Two arcs drawn from one focus of a spherical conic, to any point on the curve, and to the point where the tangent arc at the former point meets the director arc belonging to the focus, are always at right angles.

2°. The theorem given in Art. 191, 3°, is the reciprocal of Art. 192, 2°.

3°. Two fixed arcs being drawn touching a spherical conic, and any third tangent arc intersecting the two former in two

2 c

points, the arcs drawn from a focus to these two points will contain a constant angle.

4°. Theorem 6° of Art. 191 is the reciprocal of Art. 192, 4°.

5°. If tangent arcs be drawn to a spherical conic from any point in one of the director arcs, the envelope of the arc joining the points in which these tangent arcs meet two fixed tangent arcs, is a second spherical conic; the focus corresponding to the director arc in question is one of the foci of the second conic, and its director arc with respect to that curve is the arc joining the points of contact of the fixed tangent arcs.

196. The theorems arrived at in Art. 193 being reciprocated give the following results:—

1°. *If round a fixed point on the sphere, as vertex, a right spherical angle be made to turn, and if the points in which its sides meet a given spherical conic be joined by an arc, this arc will touch a second conic, of which the fixed point will be a focus; and this point will have the same polar arc in relation to the two curves.*

2°. *If a spherical right-angled triangle be inscribed in a given spherical conic, its vertex being fixed, the hypotenuse constantly passes through a fixed point.* A particular case of this is given in Art. 148, 4°.

It is easy to see, by taking a limiting position of the triangle, that *the fixed point lies on a great circle normal to the curve at the fixed vertex.*

From the first result stated in this Article, it follows that *the envelope of a spherical chord, subtending a right angle at the centre of the spherical conic, is a lesser circle.* For, since the fixed vertex has the same polar in relation to the given curve and to the envelope, and is the centre of the first, it is easy to prove (Art. 186, 5°) that it must also be the centre of the second; but it is also one of its foci; therefore the centre and a focus of the envelope coincide, which can only happen when it is a circle.

The particular theorem just proved may also be readily deduced from the polar equation of the spherical ellipse referred to its centre as pole. (See Art. 179, 2°).

Let CP and CQ be two rectangular arcs drawn from the centre, C, of the spherical conic to two points, P and Q, on the curve; let CO be a perpendicular to the arc PQ, and let the spherical angle made by CQ with the major axis be call θ. We have then

$$\frac{\cos^2\theta}{\tan^2 a} + \frac{\sin^2\theta}{\tan^2 b} = \cot^2 CQ, \text{ and } \frac{\sin^2\theta}{\tan^2 a} + \frac{\cos^2\theta}{\tan^2 b} = \cot^2 CP,$$

and therefore, by addition,

$$\cot^2 a + \cot^2 b = \cot^2 CP + \cot^2 CQ.$$

Again, COP being a right-angled spherical triangle, we have

$$\cos OCP = \tan CO \cdot \cot CP,$$

and, in like manner,

$$\cos OCQ = \tan CO \cdot \cot CQ;$$

squaring both sides of these equations, and adding, we find

$$1 = \tan^2 CO \cdot (\cot^2 CP + \cot^2 CQ),$$

or $\quad \cot^2 CO = \cot^2 CP + \cot^2 CQ = \cot^2 a + \cot^2 b.$

CO is therefore constant, which proves that PQ touches a lesser circle, whose spherical centre is C.

197. If we suppose the centre of the sphere to recede to an infinite distance upon the radius, which passes through the centre of the spherical conic, considered as a hyperbola (see Art. 184), the curve will gradually degenerate into a plane hyperbola, and the cyclic arcs will become two fixed right lines drawn through its centre. Any of the fundamental properties of the cyclic arcs already proved is sufficient to indicate that these two fixed right lines are the asymptotes of the hyperbola. In the same manner the foci and director arcs of the spherical curve will ultimately become the foci and directrices of the curve *in plano*. We are thus furnished with a method of deducing properties of the *hyperbola*, which may, of course, in certain cases be extended by the principle of continuity to the ellipse and the parabola.

In order to explain more particularly what has been said above with respect to the *asymptotes, foci, and directrices* of the hyperbola, let us, in the first place, suppose that the property of the cyclic arcs given in Art. 181, 4°, is expressed by an equation modified, in the usual way, by the introduction of any radius, r, in place of a radius equal to unity. We shall then have

$$\frac{\sin OP \cdot \sin OQ}{r^2} = \text{constant},$$

OP and OQ being perpendicular arcs from any point O of the curve to the cyclic arcs; and therefore $\sin OP \cdot \sin OQ = $ constant. Let us now suppose r to become infinite; the sines of OP and OQ will be coincident with the arcs themselves; and we shall have, in the plane figure, $OP \cdot OQ = $ constant, OP and OQ being right lines perpendicular to the limits of the cyclic arcs. Let us now draw OY and OX (in the plane figure) parallel to these limiting lines C'P and C'Q respectively, and let us represent the angle PC'Q by δ; $OQ \cdot OP$ will then be equal to $OY \cdot OX \cdot \sin^2\delta$; and therefore $OY \cdot OX = $ constant. This proves that the curve is a hyperbola, whose asymptotes are the limiting lines C'P, C'Q.

In the next place, it is evident, from what has been said in Art. 184, that the limiting positions of the foci of the spherical hyperbola are the foci of the plane hyperbola.

Finally, since the arcs between the centre of the spherical hyperbola (see Art. 184) and one focus and the corresponding vertex and director arc of the spherical ellipse, are the complements of the analogous arcs, counted from the centre of the spherical ellipse, if we call the former arcs c', a', d' respectively, we shall have (Art. 187) $\cot c' \cdot \cot d' = \cot^2 a'$, or, $\tan c' \cdot \tan d' = \tan^2 a'$. From this it follows that the director arcs of the spherical conic, in their limiting state, become the directrices of the hyperbola.

198. The following are examples of the method of deduction explained in the last Article:—

1°. Two tangents to a hyperbola, and the right line joining the points of contact, meet an asymptote in three points, the third of which bisects the distance between the first two (Art. 192, 2°).

2°. If the two sides of a variable angle, whose vertex traverses a hyperbola, pass through two fixed points on the curve, the segment intercepted between the sides of this angle upon an asymptote will be of constant length (Art. 192, 3°).

This theorem may be stated in another form:—"If the two

sides of an angle pass through two fixed points and intercept upon a given right line a segment of constant length, the vertex of this angle will generate a hyperbola, which will pass through the two fixed points, and which will have the given right line for an asymptote."

3°. *If two tangents to a hyperbola intercept on one of the asymptotes a segment of constant length, the locus of their point of concourse will be a second hyperbola. The chord joining the points of contact will envelope a third hyperbola. And the asymptote in question will also be an asymptote of the two new hyperbolas.* (Art. 192, 4°).

4°. "If in the plane of any conic section a right angle be made to turn round a fixed point as vertex, the chord subtended in the curve by the sides of this angle will touch a second conic, one of whose foci will be at the fixed point, and the corresponding directrix will be the polar of this point with relation to the given conic section."

"If the vertex of the moveable angle be the centre of the given conic, the second conic will be a concentric circle."

"If the vertex of the angle be on the given conic, the chord subtending it will constantly pass through a fixed point on the normal to the conic drawn from the vertex of the angle." (Art. 196).

5°. If tangents be drawn to any given curve of the second degree from any point in one of the directrices, the envelope of the right line joining the points in which these tangents meet two fixed tangents is another curve of the second degree; the focus corresponding to the directrix in question is one of the foci of the second curve, and its directrix with respect to that curve is the chord of contact of the two fixed tangents. (Art. 195, 5°).

ORTHOGRAPHIC PROJECTION.

199. Before concluding this Chapter, it is necessary to observe that the term *projection* is sometimes used in a more particular sense than that which we have constantly attached to it. If we suppose the centre of projection to be at infinity in a direction perpendicular to a given plane, the projection of any figure upon this plane is properly called its *orthographic* projection. In

describing this particular case, the term "orthographic" is frequently omitted; but the context is, in general, sufficient to prevent any confusion arising from this circumstance.

The orthographic projections of plane figures are possessed of certain important properties, which we shall now explain.

1°. "Any two right lines parallel to one another in the original figure are projected into two right lines, which are also parallel."

2°. "The length of the projection of a finite right line is equal to the length of the line multiplied by the cosine of the angle made by the given line with its projection." A line is evidently equal to its projection when they are parallel.

3°. *If the given figure be closed, the projected figure is also closed; and its area is equal to that of the original multiplied by the cosine of the angle between the two planes.*

The first part of this theorem is evident; the second may be proved as follows:—If we conceive the given area to be divided into an infinite number of trapezia by right lines drawn perpendicular to the intersection of the two planes, the area of the projected figure will (Prop. 1°) be divided into a corresponding system of trapezia by the projections of these lines. Now, since the area of a trapezium is equal to the rectangle under its altitude and half the sum of the parallel sides, and since the sum of the altitudes of all the trapezia in each figure is evidently the same (the altitudes being, in each case, parallel to the line of intersection of the planes), the projected area is equal to (by Prop. 2°) the given area multiplied by the cosine of the angle made by any side of one of the first set of trapezia with its projection. But this angle is the angle between the planes; and the Proposition immediately follows.

4°. *If the given figure be a circle, its projection is an ellipse, whose major axis is the projection of that diameter of the circle which is parallel to the intersection of the two planes* (and therefore equal to the diameter); *and whose minor axis equals the diameter multiplied by the cosine of the angle between the planes.*

This appears at once from Prop. 2°, combined with the following well-known property of an ellipse:—

" If on the axis major of an ellipse, as diameter, a circle be described, any perpendicular drawn from the circumference of the circle on the diameter is to the coincident perpendicular from the circumference of the ellipse, in the constant ratio of the axis major to the axis minor."

Hence it follows that any given ellipse may be regarded as the orthographic projection of a circle. This principle is very useful in deducing properties of the ellipse from those of the circle. A similar method will not apply to the hyperbola or parabola, but the properties established for the ellipse admit of being transferred (with the usual restrictions) to the other curves, by means of the principle of continuity.

5°. "When a circle is projected orthographically into an ellipse, any two rectangular diameters of the former are projected into a pair of conjugate diameters of the latter."

Because two rectangular diameters of the circle are such that each bisects the chords parallel to the other. For by Propositions 1° and 2° the projected lines will possess the same property, that is, they are conjugate diameters.

200. We shall now give some examples to illustrate the application of the principles explained in the last Article.

1°. Let CP and CQ be any two radii of a circle at right angles to one another, and CP', CQ' a second pair also at right angles (see fig.); let $P'S$ and $Q'T$ be perpendicular to the diameter PC; then we shall have the triangles CSP' and CTQ' equal in all respects. Hence we have the following property of an ellipse:—

"From the extremities of any pair of semi-conjugate diameters of an ellipse, let a pair of ordinates be drawn to a diameter of the ellipse; the two triangles so formed are equal in area."

2°. It is evident, on the same figure, that $CS^2 + TC^2 = CP'^2 = CP^2$. Hence we infer, that "If from the extremities of any pair of semi-conjugate diameters of a given ellipse two ordinates are drawn to a given diameter of the ellipse, the sum of the squares of the segments between the centre and the feet of the ordinates is equal to the square of half the given diameter, and is therefore constant." In the next figure, CP' and CQ' represent a pair of semi-conjugates, and P'S, Q'T a pair of ordinates to CA, the given diameter.

If we suppose the given diameter to coincide successively with each of the axis of the ellipse (see fig.), we shall have, by what has been proved,

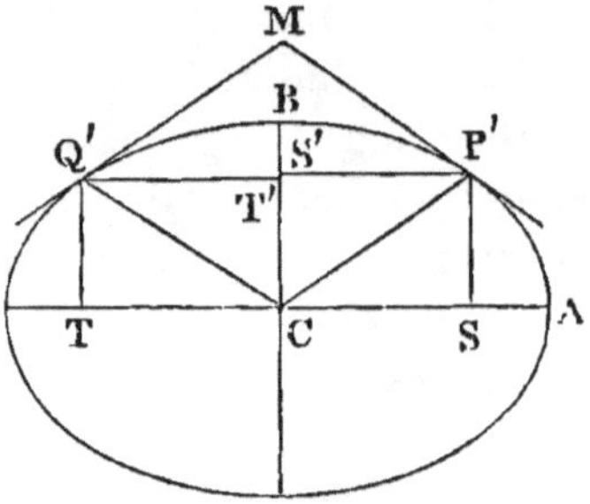

$$CS^2 + CT^2 = CA^2,$$
and $$CS'^2 + CT'^2 = CB^2,$$
and therefore, by addition,
$$CP'^2 + CQ'^2 = CA^2 + CB^2,$$

that is, *the sum of the squares of any pair of semi-conjugate diameters of an ellipse is equal to the sum of the squares of the semi-axes.*

3°. If a parallelogram of given area be circumscribed to a given circle, it is easily proved that the distances from the centre to a pair of adjacent angles are both determined; and therefore that the locus of each of these angles is a concentric circle. Hence we find that *if a parallelogram of given area be circumscribed to a given ellipse, each pair of opposite angles will lie on a similar and similarly posited concentric ellipse.*

4°. If a square be circumscribed to a circle, all its angles lie on a concentric circle, the square of whose radius is equal to twice the square of the radius of the given circle; and its area is equal to the square of the diameter. From this it appears that "if tangents be drawn to a given ellipse at the extremities of any pair of conjugate diameters, the angles of the parallelogram so made will lie on a similar and similarly posited concentric ellipse, the squares of whose semi-axes are respectively the doubles of those of the given ellipse."

It also appears that the parallelogram is of a constant area, and therefore equal to the rectangle under the axes.

It may be observed that, in this case, the two ellipses mentioned in Prop. 3° will coincide.

5°. It is evident from the figure of Prop. 1° that the areas of the segments of the circle, cut off by the chords PP′ and QQ′, are equal. Hence we have the following theorem:—"The chord joining the extremities of any two semi-diameters of an ellipse cuts off a segment equal in area to that cut off by the chord which joins the extremities of two semi-diameters, respectively, the conjugates of the former."

The triangles standing on the chords, and having a common vertex at the centre, are also evidently equal.

6°. The theorems given in Art. 83 lead to the following results:—

"If a triangle be inscribed in a given ellipse, so that one side may pass through a given point, the remaining sides intercept on the polar of that point segments which (measured from the point where the polar is cut by a diameter through the given point) contain a constant rectangle."

"If any chord be drawn through a given point in a diameter of a given ellipse, the right lines joining its extremities with one end of the diameter intercept, on a given line drawn parallel to the chords of the diameter, segments which contain a constant rectangle." (The segments are measured from the point where the parallel is cut by the diameter.)

"If any chord be drawn, as in the last enunciation, tangents at its extremities will intercept upon a tangent, drawn at one end of the diameter, segments which (measured from the point of contact) will contain a constant rectangle."

201. We shall now apply the principle of continuity to some of the results obtained in the preceding manner for the ellipse.

1°. Let us suppose, in the second figure of the last Article, CA and CB to be a pair of conjugate semi-diameters, and CP′, CQ′ to be a second pair, let the angle

$$ACP' = \theta, \ BCP' = \phi, \ ACQ' = \theta', \ BCQ' = \phi';$$

Proposition 1° of that article may then be thus expressed:—

$$CP'^2 \cdot \sin\theta \cdot \sin\phi = CQ'^2 \cdot \sin\theta' \cdot \sin\phi'.$$

Now it is known that the equation of the hyperbola, referred to a pair of conjugate diameters as axes of co-ordinates, is derived from that of the ellipse, by changing the sign of the square of the semi-diameter, which does not meet the curve. If then CP′

and CQ′ be two such semi-diameters in the figure relating to the hyperbola, which corresponds to that of the ellipse before referred to, we shall have, (observing that the angle BCQ′ also changes its sign), for the hyperbola (see fig.),

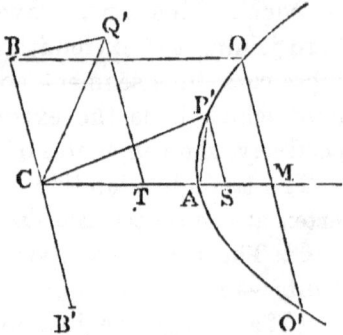

$$CP'^2 \cdot \sin\theta \cdot \sin\phi = CQ'^2 \cdot \sin\theta' \cdot \sin\phi',$$

and therefore the triangle CP′S is equal to the triangle CQ′T.

If CA be the major semi-axis, the equation will become
$$CP'^2 \cdot \sin\theta \cdot \cos\theta = \mp CQ'^2 \cdot \sin\theta' \cdot \cos\theta',$$
or
$$CP'^2 \cdot \sin 2\theta \pm CQ'^2 \cdot \sin 2\theta' = 0.$$

The upper sign being taken in the case of the ellipse, and the lower in that of the hyperbola.

2°. Retaining the same suppositions, let the angle ACB $= \omega$; the equation $CS^2 + CT^2 = CA^2$ (for the ellipse) may be written

$$\frac{CP'^2 \cdot \sin^2\phi}{\sin^2\omega} + \frac{CQ'^2 \cdot \sin^2\phi'}{\sin^2\omega} = CA^2.$$

In the case of the hyperbola, this becomes (changing the sign of CQ'^2) $CS^2 - CT^2 = CA^2$.

Again, the equation $CP'^2 + CQ'^2 =$ constant, becomes, of course, $CP'^2 - CQ'^2 =$ constant = the difference of the squares of the semi-axes.

3°. It was proved in Proposition 4° of the last Article that the locus of M, the intersection of tangents at P′ and Q′ (see the second figure of that Article), is an ellipse, similar and similarly posited to the given one, the ratio of the coincident semi-axes being $\sqrt{2} : 1$. Now MQ′ may be regarded as one of the distances *from* M to *two* points of the curve, taken parallel to the semi-diameter CP′, and equal to it. If CP′² were to become negative, one of the equal distances, MQ′, measured from M, must change its sign. The corresponding theorem for the hyperbola is then as follows:—

" If a chord, OO′, of a hyperbola (see last figure) be equal to the parallel diameter, BB′, the locus of its middle point, M, is another hyperbola similar and similarly situated to the given one, the ratio of the coincident semi-axes being as $\sqrt{2} : 1$."

The foregoing result may be readily verified. For, by a fundamental property of the curve, we have

$$MO^2 : CM^2 - CA^2 :: CB^2 : CA^2;\text{ but } MO = CB:$$

therefore $CM^2 - CA^2 = CA^2$, or $CM^2 = 2CA^2$,

and $CM : CA :: \sqrt{2} : 1$, from which the Proposition above stated follows at once.

4°. It was proved for the ellipse that the triangles whose common vertex is at C (see the second figure of the last Article), and whose bases are the chords of the arcs AP′ and BQ′, are equal in area; and therefore that

$$CA \cdot CP' \cdot \sin \theta = CB \cdot CQ' \cdot \sin \phi'.$$

Hence, for the hyperbola, we have (see last figure)

$CA \cdot CP' \cdot \sin ACP' = CB \cdot \sqrt{-1} \cdot CQ' \cdot \sqrt{-1} \cdot (-\sin BCQ')$
$= CB \cdot CQ' \cdot \sin BCQ'$, and therefore the triangle $ACP' =$ the triangle BCQ'.

5°. Since the ellipse, whether considered geometrically as a section of a cone or algebraically as a curve of the second degree, possesses an equal generality with the hyperbola, and since the parabola may be regarded as a limiting case of either (the centre being at an infinite distance), the first theorem contained in Art. 200, 6°, holds good for the hyperbola and parabola, as well as for the ellipse; and its enunciation evidently requires no further modification.

The second and third theorems given in Art. 200, 6°, are also true for the parabola and hyperbola, provided that in the case of the latter the fixed point be taken on a diameter that meets the curve.

The following is a case of the second theorem just referred to:—

"If through a given point in a diameter of a given parabola any chord be drawn, the rectangle under the ordinates of that diameter, drawn from the extremities of the chord, is constant."

202. We shall conclude this Chapter with the following

Propositions, relating to the orthographic projection of a spherical ellipse:—

1°. *The orthographic projection of a spherical ellipse upon a plane touching the sphere at the centre is an ellipse, whose semi-axes are the sines of those of the spherical ellipse.*

The polar equation of the spherical ellipse, referred to the centre as the fixed pole, is (Art. 179, 2°)

$$\cot^2 \rho = \cos^2 \theta \cdot \cot^2 a + \sin^2 \theta \cdot \cot^2 b; \text{ but } 1 = \cos^2 \theta + \sin^2 \theta;$$

we have, therefore, by addition,

$$\csc^2 \rho = \cos^2 \theta \cdot \csc^2 a + \sin^2 \theta \cdot \csc^2 b,$$

or

$$\frac{1}{\sin^2 \rho} = \frac{\cos^2 \theta}{\sin^2 a} + \frac{\sin^2 \theta}{\sin^2 b},$$

which evidently proves the Proposition.

2°. *If two spherical ellipses have the same centre and the same cyclic arcs their orthographic projections on the tangent plane to the sphere drawn at the common centre are similar and similarly situated concentric ellipses.* (Graves's translation of Chasles's Memoirs on Cones and Spherical Conics, p. 101.)

For, it was proved in Art. 181, 1°, that the sines of the angle between the cyclic arcs of a spherical ellipse $= \dfrac{\sin b}{\sin a}$: and from this the Proposition enunciated is evident by Prop. 1°.

APPENDIX.

QUESTIONS ON ELEMENTARY PLANE GEOMETRY.

1. If a square be inscribed in a triangle so that one side be coincident with the base, its side equals half the harmonic mean between the base and the perpendicular from the vertex of the triangle.

2. If the square be *exscribed*, that is, if two of its angles lie on the productions of the sides through the vertex, or the extremities of the base, its side equals the product of the base and perpendicular of the triangle divided by their difference.

3. Apply the principle of continuity (see Art. 97) to explain the connexion between the last two questions.

4. When the base of the triangle is greater than the perpendicular, the latter line is an harmonic mean between the sides of the inscribed and exscribed squares.

5. In what sense is this true, when the base is *less* than the perpendicular?

6. Of the three squares inscribable in a triangle, the greatest stands on the least side of the triangle, and the least on the greatest side.

7. The three exscribed squares follow the same order.

8. The area of a triangle equals the rectangle under the radius of the exscribed circle touching any side, and the excess of the semiperimeter above the side touched by the circle.

9. The reciprocal of the radius of the circle inscribed in a triangle equals the sum of the reciprocals of the radii of the three exscribed circles.

10. The same reciprocal also equals the sum of the reciprocals of the three perpendiculars of the triangle.

11. Given the three perpendiculars of a triangle, the triangle may be constructed.

12. Hence, given any three of the radii of the four circles which touch the three sides of a triangle, the triangle may be constructed. (Art. 6, 4° and 5°.)

13. Lines from the angles of a triangle to the points of contact of the exscribed circles with the opposite sides, meet in a point. (Art. 9, Lemma 1.)

14. The three perpendiculars of a triangle meet in a point.

15. The orthographic projection of the diagonal of a parallelogram upon a right line equals the sum or difference of the projections of two adjacent sides conterminous with the diagonal.

16. The first term of an infinite decreasing geometric progression is an harmonic mean between the sum of the series and the *sum* obtained by taking the terms alternately positive and negative. Required a geometrical proof.

17. Let AB be the base of a triangle, whose vertex is D, and let C be a point cutting the base internally, so that $m \cdot AC = n \cdot BC$; prove that $m \cdot AD^2 + n \cdot BD^2 = (m+n)(CD^2 + AC \cdot CB)$.

18. If the point C cut the base externally, so that $m \cdot AC = n \cdot BC$, prove that $m \cdot AD^2 - n \cdot BD^2 = (m-n)(CD^2 - AC \cdot CB)$.

19. Hence a point C may be found in a given right line AB (or its production), so that $m \cdot AC^2 \pm n \cdot BC^2$ shall equal a given quantity, m and n being given numbers.

20. If a transversal cut, in the points A, C, B, three right lines diverging from a point D, prove that $BC \cdot AD^2 + AC \cdot BD^2 - AB \cdot CD^2 = AB \cdot BC \cdot CA$.

21. The same relation holds good if D is any point in the right line containing A, C, B. (Art. 96.)

22. If ABCD be a quadrilateral inscribed in a circle, and O be any point, $AO^2 \cdot$ triangle $BCD + CO^2 \cdot$ triangle $ABD = BO^2 \cdot$ triangle $ACD + DO^2 \cdot$ triangle ABC. (Question 17.)

23. Given in magnitude and position the bases of a number of triangles having a common vertex; given also the sum of their areas; the locus of the vertex is a right line (Art. 72, Lemma 4). The locus may become indeterminate?

24. If some of the areas are to be subtracted from the sum of the rest, and if the remainder equals a given quantity, the locus of the common vertex is still a right line.

25. The locus of a point from which perpendiculars drawn to a number of given right lines have a given sum, is a right line. (Question 23.)

26. The locus becomes indeterminate or impossible when the given lines form an equilateral polygon.

27. How is the last result to be understood when the point is taken *without* the polygon?

28. If a quadrilateral be circumscribed to a circle, then the line joining the points of bisection of the diagonals passes through the centre. (Art. 72, Lemma 4.)

29. If the angles at the base of a triangle be bisected, and the points be joined in which the bisectors cut the sides, the portion of the joining line within the triangle has this property:—That a perpendicular on the base from any of its points equals the sum of the perpendiculars on the sides. (Quest. 24.)

30. The line mentioned in the last question meets the base in the point where it meets the external bisector of the vertical angle.

31. If a right line be drawn parallel to the base of a triangle, and its intersections with the sides be joined to two given points in the base, the locus of the intersection of the joining lines is one of two right lines. (Art. 21.)

32. The locus of the centre of a circle bisecting the circumferences of two given circles is a right line, whose distance from one of the centres equals that of the radical axis from the other. (Note to Art. 53.)

33. Describe a circle through two given points, so as to cut from a given circle an arc equal to a given arc of the same circle. (Note to Art. 62.)

34. If from a point perpendiculars be drawn on the three sides of a given triangle, the area of the triangle formed by joining the feet of the perpendiculars has a constant ratio to the rectangle under the segments of a chord of the circle circumscribed to the given triangle, drawn through the point.

35. What is the locus of the point when the feet of the three perpendiculars are in one right line?

36. When is the area of the triangle, formed by joining the feet of the perpendiculars, a maximum?

37. Given any number of points, A, B, C, &c., prove that the locus of a point O, such that $m \cdot AO^2 + n \cdot BO^2 + p \cdot CO^2 +$ &c., equals a given quantity (m, n, p, &c., being also given), is a circle. (Quest. 17.)

38. When the circle vanishes into a point (which we shall call K), the quantity $m \cdot AO^2 + n \cdot BO^2 +$ &c. (which may be expressed by the notation $\Sigma (m \cdot AO^2)$, is a minimum, and its minimum value $\Sigma (m \cdot AK^2)$ = its general value $\Sigma (m \cdot AO^2) - (m + n +$ &c.$) KO^2$.

39. If m, n, p, &c., be each equal to unity, the point K coincides

with the *centre of mean distances* of the given points A, B, C, &c. (See Lardner's Euclid, Appendix I. Section 5.)

40. If lines be drawn from any point on one of two given concentric circles to the angles of a regular polygon inscribed in the other, the sum of their squares is constant, and equal to $n \cdot (r^2 + r'^2)$, where n expresses the number of the sides of the polygon, and r and r' are the radii of the circles.

41. If a regular polygon of n sides be inscribed in a circle, whose radius is r, the sum of the squares of perpendiculars from its angles upon any diameter is equal to $\frac{1}{2} \cdot nr^2$.

42. If from a point in one side of a triangle parallels be drawn to the remaining sides, the area of the parallelogram so formed is a maximum when the side is bisected. (Art. 84, Lemma 8).

43. Hence, inscribe in a given segment of a circle a maximum rectangle.

44. If a point be taken within a triangle so as to be the centre of mean distances of the feet of perpendiculars drawn from it upon the three sides, the sum of the squares of the perpendiculars is a minimum. (Quest. 39.)

45. Given the base of a triangle and the sum of the sides, the rectangle under perpendiculars from the extremities of the base upon the external bisector of the vertical angle is constant. (Art. 80, Lemma 7.)

46. Given the base and difference of sides of a triangle, the rectangle under perpendiculars from the extremities of the base upon the internal bisector of the vertical angle is constant. (Art. 79, Lemma 6.)

47. Prove that there is no point within a triangle from which the sum of perpendiculars drawn to the three sides is a maximum or minimum.

48. The sum of the squares of the mutual distances of a system of points taken in pairs, is equal to $n \cdot$ (the sum of the squares of their distances from their centre of mean distances), n being the number of the points. (Quest. 39.)

49. Also, the square of the distance from any one of the points to the centre of mean distances equals $\frac{P}{n} - \frac{Q}{n^2}$, where P expresses the sum of the squares of the distances from the point to all the other points, and Q the sum of the squares of the mutual distances. (Quest. 48.)

50. If the mutual distances be bisected, the sum of the squares of the lines joining the points of bisection in pairs, equals $\frac{n-2}{4} \cdot Q$. (Carnot, Géométrie de Position, p. 332.)

51. The line joining two points is cut in involution by their polars, with respect to a circle, and by the circle.

52. Let the vertex of a right-angled triangle be fixed, and let its hypotenuse be a chord of a given circle, find the locus of the point of bisection of the hypotenuse.

53. The curve *inverse* (see Art. 171) to a right line is a circle.

54. The curve inverse to a circle is, in general, another circle.

55. When is the locus, inverse to a circle, not a circle? (Quest. 53.)

56. Find the locus of the pole of the hypotenuse of the right-angled triangle mentioned in Quest. 52.

57. If a point move along a given right line or circle, the locus of the *middle point* of its polar (see Art. 48, 2°), with respect to a given circle, is, *in general*, another circle.

58. If two tangents be drawn to a circle, and the points of contact be joined respectively to the extremities of a diameter, the line passing through the intersection of the joining lines and that of the tangents is perpendicular to the diameter.

59 If a variable line be drawn through a given point in the line joining the centres of two given circles, the line joining its poles with respect to the circles passes also through a fixed point.

60. When do the two fixed points coincide?

61. Draw from a given point a line so that the line joining its poles with respect to two given circles may pass through a given point.

62. Through any point in the circumference of one of two given concentric circles, let any two right lines be drawn; the line joining their poles, with respect to the other circle, touches a fixed circle.

63. Find a point within a triangle, such that lines joining it to the three angles shall divide the triangle into three equal triangles.

64. The point so found is such that the product of the perpendiculars drawn from it to the three sides of the triangle is a maximum.

65. Given the radius of a circle, find its centre, so that the area of the triangle *polar* to a given triangle, with respect to the circle (see Art. 51), shall be a minimum. (Quest. 64.)

66. Find the locus of a point within an isosceles triangle, from which a perpendicular drawn to the base is a mean proportional between perpendiculars from the same point to the sides. (Art. 46.)

67. If a line cut the sides of an isosceles triangle, so that the rectangle under the segments next the vertex is to the rectangle under the segments next the base in a constant ratio, the *envelope* of the line is a circle for one value of the constant. Prove this, and find the value. (Art. 47.)

68. If all the sides of a polygon are parallel to given lines, and if all its angles but one move on given lines, the locus of the remaining angle is a right line.

69. To inscribe in a given polygon another of the same number of sides, so that all its sides shall be parallel to given lines; or that a specified number of its sides shall be parallel to given lines, and that the rest shall pass through given points. (Art. 32.)

70. To inscribe in a given circle a polygon of a given number of sides, so that its sides may satisfy either of the conditions indicated in the last question. (Art. 48, 5°.)

71. If tangents be drawn at the extremities of a diameter of a circle, and be cut by any third tangent, the lines joining the intersections to the opposite extremities of the diameter intersect on the perpendicular to the diameter from the point of contact of the third tangent. (Art. 69, 4°.)

72. If two circles touch one another, the respective extremities of a pair of parallel diameters are *in directum* with the point of contact.

73. If a perpendicular be drawn to the diameter of a semi-circle, and a circle be described to touch the perpendicular, the given semi-circle, and another semi-circle described on one of the segments of the diameter, prove that the diameter of the touching circle equals half the harmonic mean between the segments of the diameter. (Quests. 72 and 14.)

74. When a line is cut in extreme and mean ratio, the difference of the segments equals half the harmonic mean between them.

75. Given in position the vertex of a right-angled triangle; given also the sum of the squares of the reciprocals of its sides; the envelope of the hypotenuse is a circle.

76. Given the vertical angle of a triangle, and the bisector of the angle, prove that the sum of the reciprocals of the containing sides is constant.

77. If a chord be drawn through a given point, cutting a given circle, and tangents be drawn at the points of intersection, the sum or difference of the reciprocals of perpendiculars from the given point upon the tangents is constant.

78. Given four points, A, B, C, **D**, in a right line, the locus of a point at which the angle subtended by A and B equals that subtended by C and D, is a circle. (Art 34, Lemma 3.)

79. If a quadrilateral be inscribed in another (plane or "gauche"), so that one pair of its opposite sides shall intersect on one diagonal of the second quadrilateral, its other pair of opposite sides will intersect on the other diagonal.

80. Given in position a point and a right line, find the locus of a point the square of whose distance from the given point shall have a given ratio to the rectangle, under a given line, and the distance from the point to the line given in position. (Art. 61.)

81. Find the locus of the centre of a circle, cutting one given circle orthogonally, and bisecting the circumference of another given circle.

82. The centre of a circle touching three given circles may be found by drawing a common tangent to two given circles. (Art. 81.)

83. If from a given point, O, a transversal be drawn cutting a number of given circles in A, A', B, B', &c., and if tangents at these points intersect a given right line, also drawn from O, in X, X', Y, Y', &c., then, $\frac{1}{OX} + \frac{1}{OX'} + \frac{1}{OY} + \frac{1}{OY'} +$ &c., is constant. (Art. 48, 1°.)

84. Cut a given line internally so that the square of one segment may equal the rectangle under the other segment and a given line.

85. Solve the corresponding problem when the line is cut externally, and find the limits within which the problem is possible when the greater segment is that whose square is taken.

86. Cut a given line harmonically, so that the middle part may equal a given line, and find when the solution becomes impossible.

87. Given a circle and a right line in position, a point may be found such that any chord being drawn through it, and perpendiculars drawn from its extremities upon the given line, the perpendiculars shall be to one another in the duplicate ratio of the segment of the chord. (Art. 60, 1°.) (The line and circle are supposed not to intersect.)

88. If every pair of three triangles be *copolar* (see Art. 24), in such a manner that the three lines joining corresponding angles are the same for all, the three corresponding *axes* meet in a point.

89. Reciprocate the theorem last given.

90. Given all the sides of a polygon but one, prove that its area is a maximum when it is inscribable in a semi-circle, of which the remaining side is the diameter.

91. Given the length of a circular arc, prove that the area of the corresponding segment is a maximum when it is a semi-circle. (Quest. 90.)

92. If three chords of a circle be drawn from the same point on the circumference, and if on these, as diameters, three circles be described, the intersections of every pair of the circles (excluding the given point) are *in directum*. (Quest. 35.)

93. Given the vertical angle of a triangle, and the sum of the containing sides, find the locus of a point cutting the base in a given ratio.

94. The case where the *difference* of the side is taken in place of the sum may be deduced from the last by the principle of continuity?

95. From the anharmonic property of four points on a circle it may be proved that the rectangle under the diagonals of an inscribed quadrilateral equals the sum of the rectangles under the opposite sides. (Art. 26.)

96. If a regular polygon of an *odd* number of sides be inscribed in a circle, and the angles be joined to any point on the circumstance, the sums of the alternate sets of joining lines are equal. (Quest. 95.)

97. If three transversals be drawn from the angles of a triangle ABC, meeting in a point O, and cutting the opposite sides in A', B', C', then, $\dfrac{OA'}{AA'} + \dfrac{OB'}{BB'} + \dfrac{OC'}{CC'} = 1.$

98. In a given circle inscribe a triangle, so that two sides may pass through given points, and that the third be a maximum or minimum. (Art. 66, 3°.)

99. If three lines be in harmonic proportion, the rectangle under the extremes exceeds the square of the mean by the rectangle under the differences between the mean and each extreme. (Art. 4.)

100. Given in magnitude and position the base of a triangle whose vertex is on a given line, construct the triangle so that the line joining the vertex to the middle of the base may be a mean proportional between the sides. (Note to Art. 66, 3°.)

101. Draw a chord of a given circle parallel to a given line, so that the triangle whose base is the chord and whose vertex is at a given point may be a maximum. (Quest. 42.)

102. If a, b, c be three magnitudes in arithmetic proportion, and if b, c, d be in harmonic proportion, then $a : b :: c : d$.

103. From a given point in the produced base of a triangle, it is required to draw a line cutting the sides, so that the difference of perpendiculars drawn to the base from the points of intersection may be a maximum. (Art. 84, Lemma 8.)

104. If a *direct* and a *transverse* common tangent (see Art. 65, 1°,) be drawn to two circles, the difference of the squares of the parts intercepted by the points of contact equals the rectangle under the diameters.

105. Find a point within a triangle such that the continued product of the segments of three lines parallel to the sides, and drawn through the point, shall be a maximum. (Quest. 64.)

106. Given the base and sum of sides of a triangle, prove that the portion of one side, between the vertex and the foot of a perpendicular drawn to the side from the extremity of the bisector of the vertical angle, is constant.

107. From a given point in the produced base of a triangle, to draw a line cutting the sides, so that the area of the quadrilateral completed by joining the points of intersection to two given points in the base may be a maximum. (Art. 84, Lemma 8.)

108. Given in position a circle and a right line, two points can be found, with respect to either of which, taken as *origin*, the circle and line shall be mutually *inverse*. (See Art. 171.)

109. Given the base, the vertical angle, and either bisector of the vertical angle, construct the triangle. (Quest. 108.)

110. The lines joining the feet of the perpendiculars of a triangle form a new triangle, the centres of whose inscribed and exscribed circles are the angular points of the original triangle, and the intersection of its perpendiculars. (Quest. 14.)

111. A circle described through the feet of the three perpendiculars of a triangle bisects the three sides, and also bisects the distances between the angles of the triangle and the intersection of its perpendiculars.

112. Given the base and vertical angle of a triangle, prove that the locus of the centre of the circle passing through the centres of the three exscribed circles is a circle. (Quest. 111.)

113. Describe a circle passing through a given point, and touching a given circle and a given right line. (Quests. 72 and 108.)

114. To find a point in a given right line, so that lines joining it to two given points may make with the given line angles having a given difference.

115. Describe a circle through two given points to cut a given right line, so that the angle contained in either segment of the circle cut off by the line shall equal a given angle. (Quest. 114.)

QUESTIONS RELATING TO CIRCLES ON THE SPHERE.

116. An angle in a semi-circle is a right angle. What is the analogous theorem on the sphere?

117. If a spherical quadrilateral be inscribed in a lesser circle, the sum of one pair of opposite angles equals that of the other pair.

118. If a spherical quadrilateral be circumscribed to a lesser circle, the sum of one pair of opposite sides equals half the perimeter of the quadrilateral.

119. Given the base of a spherical triangle, and the difference between the verticle angle and the sum of the base angles, the locus of the vertex is a lesser circle passing through the extremities of the base.

120. If a variable great circle cut a given great circle at a given angle, what is the locus of its pole?

121. Given two sides of a spherical triangle, prove that its area is a maximum when the contained angle equals the sum of the other angles of the triangle. (Art. 148, 2°.)

122. If (in the last question) a and b be the given sides, and c the third side, then, $\cos \frac{1}{2} c = \sqrt{\{\frac{1}{2}(\cos a + \cos b)\}}$. (Art. 148, 1°.)

123. Given the base of a spherical triangle and the difference of the cosines of the sides, find the locus of the vertex. (Art. 148, 1°.)

124. Given the base of a spherical triangle, and the value of $m \cdot \cos a + n \cdot \cos b$, where a and b are the sides, and m and n are given numbers, prove the locus of the vertex to be a lesser circle, whose spherical centre is found by cutting the base into two segments, x and y, such that $m \cdot \sin x = n \cdot \sin y$.

125. Hence, find the locus of the vertex of the triangle, when the base is given, and $m \cdot \cos a - n \cdot \cos b$.

126. Given the value of $m \cdot \cos a + n \cdot \cos b + p \cdot \cos c +$ &c., where $m, n, p,$ &c., are given numbers, and $a, b, c,$ &c., are the arcs joining given points to a variable point, O, on the sphere, the locus of O is a circle.

127. Given all the sides of a spherical polygon except one, what is the property of the remaining side when the area is a maximum? (Quests. 121 and 116.)

128. If a, b, c, d be the sides of a spherical quadrilateral, whose diagonals are x and y, prove that the cosine of the angle made by the diagonals equals $\dfrac{\cos a \cdot \cos c - \cos b \cdot \cos d}{\sin x \cdot \sin y}$.

129. If a and b be the sides of a right-angled spherical triangle, and x and y the segments of the hypotenuse made by a perpendicular from the vertex, $\dfrac{\tan^2 a}{\tan^2 b} = \dfrac{\tan x}{\tan y}$, and $\dfrac{\sin^2 a}{\sin^2 b} = \dfrac{\sin 2x}{\sin 2y}$.

130. Given the base and sum of sides of a spherical triangle, find the locus of the point where the external bisector of the vertical angle cuts the perpendicular to one side drawn at the extremity of the base. (Art. 148, 3°.)

131. Given the base of a spherical triangle and the two bisectors of the vertical angle, construct the triangle. (Art. 123.)

132. Given the length of an arc of a lesser circle on a given sphere, prove that the spherical area, contained between it and an arc of a great circle joining its extremities, is a maximum when this latter arc passes through the spherical centre of the lesser circle. (Quest. 127.)

133. Given in position the vertical angle of a spherical triangle and a point through which the base passes, prove that its area is a maximum or minimum when the base is bisected by the point.

134. What is the reciprocal of the last theorem?

135. If the vertical angle of a spherical triangle equals the sum of the base angles, the square of the tangent of half the perpendicular from the vertex on the base equals the product of the tangents of the halves of the segments of the base. (Art. 133, Lemma 12.)

136. Also, the square of the tangent of the perpendicular is to the product of the sines of the segments of the base in a constant artio.

137. If perpendiculars be drawn from a given point, to tangent arcs at the extremities of a spherical chord, drawn through a given point, and cutting a given lesser circle, the sum or difference of their cosecants is constant.

138. If θ be the angle made by the base of a spherical triangle with the arc joining the middle points of the sides, prove

$$\tan \theta = \frac{\tan \tfrac{1}{2} \text{ area}}{\sin \tfrac{1}{2} \text{ base}}.$$

139. If p, p', p'', &c. be perpendicular arcs from a point on a number of given great circles, and if $m \cdot \sin p + m' \cdot \sin p' + m'' \cdot \sin p'' +$ &c., equals a given quantity, m, m', m'', &c., being given numbers, find the locus of the point. (Quest. 126.)

140. Given the base of a spherical triangle and the ratio of the tangents of the base angles, the locus of the vertex is a great circle.

141. What is the reciprocal of this property?

142. In a right-angled spherical triangle, the difference of the angles at the hypotenuse is less than a right angle, and their sum less than three right angles.

143. If m be the arc joining the middle points of the sides of a spherical triangle, then, $\cos m = \cos \tfrac{1}{2} \text{ area} \cdot \cos \tfrac{1}{2} \text{ base}$.

144. Given an acute angle at the hypotenuse of a right-angled spherical triangle, the difference of the hypotenuse and side containing the angle is a maximum when their sum is 90° or 270°.

145. Given the four sides and the two diagonals of a spherical quadrilateral, find the expression for the cosine of the arc joining the middle points of the diagonals. (Art. 148, 1°.)

146. What does this relation become when the radius of the sphere is supposed to be made infinite?

147. Deduce the expression for the area of a plane triangle in terms of its sides from the analogous expression relating to a spherical triangle.

148. If the perpendicular of a spherical triangle fall within the base, the tangent of half the difference of the sides is a mean proportional between the tangents of the halves of the differences of the segments of the base made by the perpendicular and by the bisector of the vertical angle respectively.

149. What is the analogous theorem *in plano?* and how must it be modified when the perpendicular falls without the base of the triangle?

150. If θ be the angle between the perpendicular of a spherical triangle and the bisector of the vertical angle, prove

$$\tan \theta = \tan \tfrac{1}{2}(A - B) \cdot \frac{\cos \tfrac{1}{2}(a-b)}{\cos \tfrac{1}{2}(a+b)},$$

a and b being the two sides, and A and B the opposite angles.

151. What proposition in plane geometry corresponds to the last theorem?

152. If a, b, c be the sides of a spherical triangle, and A' the angle made by the chords of b and c, prove

$$\cos A' = \frac{1 + \cos a - \cos b - \cos c}{4 \sin \tfrac{1}{2}b \cdot \sin \tfrac{1}{2}c}.$$

153. If a point be taken on each side of a spherical triangle, such that the products of the cosines of the alternate segments are equal, perpendiculars drawn to the sides at the three points will meet in a point. (Art. 148, 3°.)

154. If from any point on the sphere perpendiculars be drawn to the sides of a spherical triangle, and their feet be joined by arcs, perpendiculars on these arcs from the angles of the original triangle meet in a point.

155. If the two points mentioned in the last question be joined to one of the angles of the original triangle, the joining arcs make equal angles with the sides containing the angle. (Art. 109.)

156. Also, the products of the sines of arcs drawn from the two points perpendicular to each side of the triangle are equal.

157. Given the base of a spherical triangle, and the value of the function N (see Art. 129), find the locus of the vertex.

158. Given the base of a spherical triangle equal to a quadrant; given also the sum of the squares of the cotangents of the base angles; find the locus of the vertex.

159. Given the base and the area of a spherical triangle, find the locus of the middle point of one of the sides. (Art. 148, 2°.)

160. If three arcs be drawn from the angles of a spherical triangle,

ABC, meeting in a point O, and cutting the opposite sides in A', B', C', respectively, then

$$\frac{\tan OA'}{\tan OA' + \tan OA} + \frac{\tan OB'}{\tan OB' + \tan OB} + \frac{\tan OC'}{\tan OC' + \tan OC} = 1.$$

(Quest. 97.)

161. If a polygon be formed on the sphere by any curves not great circles, its area is less than that of a polygon whose sides are arcs of great circles, and the sum of whose angles equals that of the former, by the sum of the arcs *polar* to the sides of the former polygon. (Art. 147.)

162. The area of a triangle formed by three lesser circles, whose spherical radii are r, r', r'', is less than the area of an equi-angular spherical triangle (in the usual sense), by $a \cdot \cot r + b \cdot \cot r' + c \cdot \cot r''$, a, b, c, being the arcs of the three lesser circles. (Art. 145, 3°.)

163. The area of the *lune* contained between two lesser circles, whose spherical radii are r and r', equals $2\theta - a \cdot \cot r - b \cdot \cot r'$, θ being the angle made by the circles, and a and b the arcs forming the lune.

MISCELLANEOUS QUESTIONS ON THE FOREGOING SUBJECTS.

164. If in Art. 70, 6°, the number $k = 1$, the line QP is cut by MN in the ratio of AM to BM, and the line MN is cut by QP in the ratio of AQ to DQ.

165. If on two given right lines (which need not be in the same plane), two portions be taken, measured from given points, and in a given ratio, the locus of a point, cutting the line joining their extremities in a given ratio, is a right line.

166. If a plane polygon be inscribed in a given one of the same number of sides, so that all its sides but one are parallel to given lines, the locus of a point cutting the remaining side in a given ratio is a right line.

167. Given two points in the diameter of a semi-circle at different sides of the centre; prove that the rectangle under their distances from a point in the circumference is a maximum when the distances are in the subduplicate ratio of the distances from the given points to the centre of the circle. (Quest. 17.)

168. Given three right lines *in space*, prove that a right line drawn from any point of one may rest on the other two.

169. If four right lines rest on three others given *in space*, the anharmonic ratios of the three systems of four points are equal. (Art. 116.)

170. If in Art. 73, 1°, the given right lines meet in a point, the

locus passes through the same point. Prove this, and find the reciprocal.

171. The results given in Questions 37, 38, 39 hold good when the given points are *not in the same plane*, a sphere being substituted in place of a circle.

172. The results given in Questions 48, 49, 50, also hold good when the given points do not lie in the same plane.

173. If three alternate sides of a plane or spherical hexagon meet in a point, and if the remaining three also meet in a point, the three diagonals joining opposite vertices of the hexagon will meet in a point.

174. Given the base of a spherical triangle and the arc joining the vertex to a given point in the base, construct the triangle so that the product of the cosines of the sides may be a maximum.

175. If the entire surface of a sphere be divided into polygons of the same number of sides, the number of the polygons added to the number of distinct corners is equal to the number of distinct sides added to 2.

176. If the circumference of a lesser circle of a sphere be divided into an odd number of equal parts, and arcs be drawn to the points of section from any point on the circumference, the sum of the sines of the halves of one alternate set is equal to the sum of the sines of the halves of the other alternate set.

177. If the circumference of a lesser circle be divided into n equal parts, the sum of the squares of the sines of the halves of arcs joining the points of section to any point on the circumference is equal to $\frac{n}{2}$. the square of the sine of the spherical radius.

178. Given three points, A, B, C, on a sphere, if a fourth point, O, be so taken that m . the arc AO $+ n$. the arc BO $+ p$. the arc CO be a minimum, m, n, and p being given numbers, then,

$$\frac{\sin BOC}{m} = \frac{\sin AOC}{n} = \frac{\sin AOB}{p}.$$

179. If from any point on the surface of a sphere three chords be drawn mutually at right angles, the sum of their squares equals the square of the diameter.

180. The distances of any two points from the centre of a sphere are to one another as their distances from the alternate polar planes of the points. (Art. 45.)

181. Find the equation connecting the cosines of the six arcs which join four points on the surface of a sphere. (See Carnot, Géométrie de Position, p. 407.)

182. Find the corresponding equation connecting the distances between every pair of four points lying in the same plane. (Carnot, Géométrie de Position, p. 388.)

183. If from a given point O a transversal be drawn cutting a number of given spheres in A, A', B, B', C, C', &c., and if on the transversal a portion OX be taken, such that
$$\frac{1}{OX} = \frac{1}{OA} + \frac{1}{OA'} + \frac{1}{OB} + \frac{1}{OB'} + \&c.,$$
the locus of the point X is a plane.

184. If tangent planes be drawn to the spheres at the points A, A', B, B', &c., the sum of the reciprocals of the segments (measured from O) which they cut off from a given right line passing through O is constant.

185. Given the base of a plane triangle and the difference of the base angles, prove that the locus of the vertex is an equilateral hyperbola of which the base is a diameter, and the conjugate of which makes with the base an angle equal to the given difference.

186. Given in position the circular base of a cone; given also a plane to which the subcontrary sections are parallel; prove that the locus of the vertex of the cone is an equilateral hyberbola. (Quest. 185.)

187. If a chord of a prolate surface of revolution of the second degree be drawn through a given point, the product of the tangents of the halves of the angles subtended at one of the foci by the segments of the chord is constant.

188. The product mentioned in the last question will remain constant if the point move on another surface of revolution of the second degree, having the focus and its corresponding director plane in common with the original surface. (By the director plane is meant the plane described by the directrix of the curve which generates the surface of revolution.)

189. If a right cone, whose vertex is one focus of a prolate surface of revolution of the second degree, have a constant vertical angle, the plane of its intersection with the surface (see Art. 168, 1°) touches a second prolate surface, and the vertex of the circumscribed cone whose curve of contact (see Art. 168, 2°) is the curve of intersection describes a third prolate surface; and these three surfaces have one focus and the corresponding director plane in common.

190. The right line joining the vertices of the two cones mentioned in the last question passes through the point of contact of the plane with its "enveloping" surface.

191. The orthographic projection of a transverse section of a paraboloid of revolution upon a plane perpendicular to the axis is a circle.

192. Given an ellipse and a point in its plane, a right line may be found such that the rectangle under perpendiculars on it from the extremities of any chord of the ellipse drawn through the point shall be constant. (Art. 60, 2°.)

193. If a number of ellipses be inscribed in a quadrilateral, the locus of their centres is the right line which joins the middle points of the diagonals. (Art. 72, Lemma 4.)

194. Find the locus of the centre of an ellipse touching three given right lines, one of the points of contact being also given. (Quest. 193.)

195. The locus of the centre of a sphere rolling on two rollers which meet one another is an ellipse.

196. Any three pairs of conjugate diameters of an ellipse form a pencil in involution.

197. If two triangles be circumscribed to an ellipse, the six vertices lie on a conic section.

198. If a triangle be inscribed in an ellipse, so that two sides are parallel to given right lines, the envelope of the third side is a similar ellipse.

199. The area of an ellipse circumscribed to a given triangle is a minimum when it is to that of the triangle :: $4\pi : 3\sqrt{3}$.

200. If a sphere be inscribed in a right cone, and be touched by any plane, the point of contact is one focus of the section made by the plane, and the corresponding directrix is the line of intersection of the same plane with that of the circle of contact of the cone and sphere.

201. If a plane be drawn cutting the circle of contact mentioned in the last question, we shall have a conic section and an inscribed circle; prove that a tangent drawn to the circle from any point on the conic is equal to the excentricity of the conic multiplied into the perpendicular from the point upon the line of intersection of the plane of the conic and the plane of the circle of contact. (Quest. 200.)

202. If two given circles have double contact *internally* with a given conic section, the sum or difference of tangents drawn to the circles from any point on the conic is constant.

203. How must the result given in Question 201 be modified in the case of a circle which has double contact *externally* with a conic section?

204. Is the result given in Question 202 still true, when the contacts are *external*?

205. The results stated in Question 200 remain true when in place of a right cone we substitute any prolate surface of revolution of the second degree.

206. If a plane be drawn through one focus of a prolate surface

the section of the surface made by it is a curve of the second degree, one of whose foci is the focus of the surface, the corresponding directrix being, moreover, the intersection of the plane with the director plane of the surface.

207. Given in magnitude and position a section of a right cone, the locus of the vertex of the cone is another conic section, whose plane is perpendicular to that of the former; and the extremities of the major axis in each of the conics are the foci of the other.

208. If two planes be drawn through the *focal lines* (see Art. 189) of an oblique cone, and through any side of the cone, they make equal angles with the plane touching the cone along that side. (Art. 181, 2°.)

209. Conversely, if two right lines be drawn from the vertex of a cone, and if planes drawn through them and any side whatever of the cone make equal angles with the plane touching the cone along that side, these right lines are the focal lines of the cone.

210. If two cones have the same focal lines and cut one another, their intersection consists of four right lines, each of which is such that planes touching the cones along this line are at right angles. (Art. 185.)

211. If from a point assumed upon a focal line of an oblique cone perpendiculars be let fall upon the tangent planes to this cone, their feet will be upon a circle, the plane of which will be perpendicular to the second focal line of the cone. (Chasles's Memoir on Cones, Art. 27.)

212. The locus of a point whose perpendicular distances from a fixed right line and plane are in a given ratio is an oblique cone, of which the given right line is a focal line. (Art. 191, 2°.)

213. If from the foci of a spherical conic arcs be drawn perpendicular to the arcs touching the curve, their respective points of intersection with these tangent arcs will be upon a second spherical conic, which will have a double contact with the given one, and whose cyclic arcs will be in the planes perpendicular to the radii of the sphere which pass through the two foci of the given conic. (Quest. 211.)

214. If arcs be drawn from any point of a spherical ellipse to the foci, the ratio or product of the tangents of the halves of the angles made by these arcs with the major axis is constant, according as the angles are measured in the same or opposite directions. (Art. 181, 1°.)

215. If a normal arc be drawn at a point of a spherical conic, and from its intersection with the major axis a perpendicular arc be drawn to a focal radius vector drawn to the point, the portion of the radius vector between the perpendicular and the curve is constant.

216. The tangents of the portions of a normal arc to a spherical

conic intercepted between the curve and the two axes are to one another in the duplicate ratio of the tangents of the semi-axes.

217. Given the base of a spherical triangle in which the tangent of half the bisector of the base is a mean proportional between the tangents of the halves of the sides, the locus of the vertex is a spherical hyperbola, whose foci are the extremities of the base.

218. Given the base of a spherical triangle, and the ratio of the sine of one side to the cosine of the other, find the locus of the vertex. (Art. 191, 2°.)

219. Find the radius of a circle *in plano*, whose area shall equal the spherical area of a given lesser circle of a sphere. (Art. 146.)

220. Given the vertical angle of a spherical triangle and its bisector, prove that the sum of the cotangents of the sides containing the angle is constant. (Quest. 76.)

221. Let two tangent arcs be drawn to a spherical conic, and let the bisector of the angle made by them be made to cut the arc joining the points of contact; if through the latter point of intersection a spherical chord of the conic be drawn, the angle subtended by the segments of the chord at the intersection of the tangent arcs are equal.

222. Given one focus of a spherical ellipse and two points on the curve, find the locus of the other focus.

223. Given one focus of a spherical ellipse and two tangent arcs, in position, find the locus of the other focus.

224. If a tangent arc subtend a constant angle at one of the foci of a given spherical conic, what is, in general, the locus of the point from which it is drawn?

225. What is the locus when the constant angle is a right angle?

226. Given in position one focus of a spherical conic and two tangent arcs, prove that the arc joining the points of contact passes through a fixed point.

227. An arc cutting a system of spherical conics which have one focus and the corresponding director arc in common gives a system of points in spherical involution.

228. Let a great circle touch a given spherical conic, and let two points be taken on it so as to subtend a constant angle at one of the foci; if one of the points move on a given tangent arc, what is the locus of the other?

229. If a number of lesser circles, having a common spherical centre, be projected orthographically upon the plane of a given great circle, the locus of the extremities of the major axes of the ellipses so formed is an ellipse.

230. Prove that the locus of the foci of the elliptical projections mentioned in the last question is a circle.

231. Given a conic section, and a right line in the same plane, *but not meeting the conic*, the conic may be projected into a circle, and the right line, at the same time, be projected to an infinite distance. (See Poncelet, Traité des Propriétés Projectives, Art. 110, or Salmon's Conic Sections, Art. 350.)

232. Projective properties of two concentric circles may be transferred to two plane or spherical conics having double contact, by the aid of the principle of continuity. (Quest. 231.)

233. If through the centre of a hyperbola an ellipse be described having for foci the points where a tangent to the hyperbola cuts the asymptotes, the ellipse touches the minor axis of the hyperbola.

234. If a normal be drawn to an ellipse at any point, and two portions be measured on it from the point, in opposite directions and equal to the semi-diameter conjugate to the point, an ellipse described with the extremities of these portions as foci, and passing through the centre of the given ellipse, will touch its minor axis.

235. Explain the connexion between the last two questions by the principle of continuity. (See Chasles, Aperçu Historique, p. 361.)

236. If the equation of a *hyperbola* referred to given axes of co-ordinates, which are inclined at an angle, ω, be

$$Ax^2 + Bxy + Cy^2 + Dx + Ey + F = 0,$$

the tangent of the angle made by the asymptotes is equal to

$$\frac{\sqrt{(B^2 - 4AC)} \cdot \sin \omega}{A + C - B \cdot \cos \omega}$$

237. Hence deduce the condition of *similarity* of *any* two curves of the second degree, whose equations are given with respect to any given axes of co-ordinates.

INDEX.

	PAGE.
ANHARMONIC RATIO,	
Of a plane pencil,	14
Of a spherical pencil,	119
Of four planes,	117
Of four points in a right line,	13
on a great circle,	121
on a lesser circle,	128
on a circle *in plano*,	19
on a plane conic,	183
on a spherical conic,	183
BRIANCHON'S THEOREM,	38
CARNOT'S THEOREM,	183
CO-AXIAL TRIANGLES,	19
CONE.	
Right,	110
Oblique,	110
Base of,	110
Sides of,	110
Anharmonic property of,	127
Principal section *of an oblique*,	115
Subcontrary sections *of an oblique*,	116
Cyclic planes *of an oblique*,	169
Focal lines *of an oblique*,	188
CONTINGENT RELATIONS.	
Principle of,	103
CONTINUITY.	
Principle of,	96

	PAGE.
COPOLAR TRIANGLES,	19
CORRELATIVE FIGURES,	99
CORRESPONDING POINTS.	
Directly,	60
Inversely,	61
ENVELOPE,	92
EXSCRIBED CIRCLE,	6
"GAUCHE" POLYGON,	72
GRAPHICAL PROPERTIES,	113
GRAVES'S THEOREM,	179
HARMONIC,	
Proportion,	1
System of Planes,	118
Conjugate points,	2
legs,	8
planes,	118
Pencils " in plano,"	7
on the sphere,	121
HARMONICALLY.	
Right line cut,	2
Arc cut,	122
IDEAL CHORD,	52
INVERSE CURVES AND SURFACES,	168

INDEX.

INVOLUTION,

In plano.
    Six points in, ... 28
    Pencil in, ... 30
    System of points in, ... 28
    Centre of a system of points in, 28
    Foci of a system of points in, . 29

On the sphere.
    Six points in, . . , . . 129
    Pencil in, ... 131
    System of points in, ... 129
    Centres of a system of points in, 130
    Foci of a system of points in, . 130

LIMITING POINTS,

    Of a system of circles having a common radical axis *in plano*, 53
    Of a system of circles having a common radical axis, on the sphere, ... 152

METRICAL PROPERTIES, . . 114

ORTHOGONALLY.

    Circles intersecting, ... 4

PASCAL'S

    Theorem, ... 20
    Line, ... 20

POLAR LINES,

    In relation to a sphere, . . . 161

POLE AND POLAR,

    In relation to a lesser circle, . 133
    a circle *in plano*, ... 32
    a plane conic, ... 184
    a spherical conic, ... 184

POLE AND POLAR PLANE,

    In relation to a sphere, . . . 159

PROJECTION.

    Stereographic, ... 165
    Orthographic, ... 197
    Centre of, ... 100
    Surface of, ... 109

QUADRILATERAL.

    Complete, ... 11
    Third diagonal of, ... 10

RADICAL AXIS,

    Of two circles *in plano*, . . . 52
    Of two lesser circles, ... 152
    Common to a system of circles *in plano*, ... 53
    Common to a system of lesser circles, ... 152

RADICAL CENTRE,

    In plano, ... 57
    On the sphere, ... 153

RECIPROCAL,

    Of a line, ... 4

RECIPROCALLY POLAR,

    Surfaces, ... 162
    Plane polygons, ... 38
    curves, ... 39
    Spherical polygons, . . . 142
    curves, ... 142

RECIPROCATION,

    In plano, ... 37
    On the sphere, general method, 140
    particular method, . . . 141

SIMILITUDE.

    Axes of, "in plano," 64
    on the sphere, ... 158
    Centres of, "in plano," . . . 59
    on the sphere, ... 155

SPHERICAL CENTRE,

    Of a lesser circle, ... 113

SPHERICAL CONICS.

    Formation of, ... 173
    Cyclic arcs of, ... 174
    Director arcs of, ... 186

SPHERICAL ELLIPSE.

    Formation of, ... 173

SPHERICAL ELLIPSE—*continued.*

  Centre of, ... 175
  Foci of, ... 177
  Semi-axes of, ... 174

SPHERICAL HYPERBOLA.

  Formation of, ... 181
  Centre of, ... 182
  Foci of, ... 181

SPHERICAL RADIUS,

  Of a lesser circle, ... 110

STEINER'S THEOREM, . . 25

SUPPLEMENTARY CONES, . . 143

SUPPLEMENTARY CURVES . 142

SUPPLEMENTARY SPHERICAL
  ELLIPSE, ... 177

TOWNSEND'S APPLICATION OF
  ANHARMONIC PROPER-
  TIES, ... 24

TRANSVERSALS.

  Fundamental propositions relating to, 9

THE END.

www.ingramcontent.com/pod-product-compliance
Lightning Source LLC
Chambersburg PA
CBHW021813230426
43669CB00008B/738